The Quantum Particle Illusion

Conceptual Quantum Mechanics

The Quantum Particle Illusion

Conceptual Quantum Mechanics

Gerald E Marsh

World Scientific

NEW JERSEY · LONDON · SINGAPORE · BEIJING · SHANGHAI · HONG KONG · TAIPEI · CHENNAI · TOKYO

Published by

World Scientific Publishing Co. Pte. Ltd.

5 Toh Tuck Link, Singapore 596224

USA office: 27 Warren Street, Suite 401-402, Hackensack, NJ 07601

UK office: 57 Shelton Street, Covent Garden, London WC2H 9HE

Library of Congress Control Number: 2021049382

British Library Cataloguing-in-Publication Data
A catalogue record for this book is available from the British Library.

THE QUANTUM PARTICLE ILLUSION
Conceptual Quantum Mechanics

ISBN 978-981-124-822-1 (hardcover)
ISBN 978-981-124-823-8 (ebook for institutions)
ISBN 978-981-124-824-5 (ebook for individuals)

For any available supplementary material, please visit
https://www.worldscientific.com/worldscibooks/10.1142/12583#t=suppl

This book is dedicated to my wife Evelyn without whom I would not and could not be who I am today. We married very early and had the privilege of traveling together through the harmony of life.

Preface

The topic of the foundations of quantum mechanics has a vast and very interesting literature and a great many papers and books have been written about the difficulties of interpreting quantum mechanics, in general, and the wavefunction, in particular. I will not make any attempt to review this literature since it stands on its own. One of the relatively most active modern periods was in the 1960s and 1970s and especially from 1973 to around 1979.

Quantum mechanics can only present the probability of a particle being found in any given region of spacetime. This raises the question of whether the position of a discrete particle can in principle only be predicted statistically by quantum mechanics or, whether the formalism of quantum mechanics applies not to a single particle, but rather to an ensemble of identical systems. The various approaches to interpreting the mathematics of quantum mechanics are motivated by somehow attempting to maintain our ideas about a classical particle in the context of the quantum world as found in nature.

This has led to great confusion not only among those approaching quantum mechanics for the first time, but also for those who have used the formalism to perform highly successful calculations for many years. In the end, it is the historical approach to the teaching of quantum mechanics that could be the root of the problem. Those brave enough to try and understand the conceptual foundations of quantum mechanics when being taught from this perspective, and often discouraged from doing so by the apocryphal advice given by many physicists to "ignore the issue and just calculate", are generally left in a state of cognitive dissidence well expressed by the adaptation of a cartoon drawn by an anonymous artist that appeared, in a completely different context, in a government sponsored report from the mid-1980s, shown below.

A poor soul who tries to understand the foundations of quantum mechanics after being taught the subject using an historical approach.

It is hoped that this book can relieve some of the dismay, frustration, and confusion so well expressed by this cartoon.

Gerald E. Marsh

Contents

Introduction

The statistical interpretation of the wavefunction, ψ, is due to Max Born who, in his 1954 Nobel prize acceptance speech, ascribed his inspiration for the statistical interpretation to an idea of Einstein's. Here is the relevant quote from Born's speech: "He had tried to make the duality of particles — light quanta or photons — and waves comprehensible by interpreting the square of the optical wave amplitudes as probability density for the occurrence of photons. This concept could at once be carried over to the ψ-function: $|\psi|^2$ ought to represent the probability density for electrons (or other particles)."

Einstein, in a December 1926 letter to Max Born, speaking of the "secret of the Old One," said that he was "convinced that He does not throw dice." And when Philipp Franck pointed out to Einstein, around 1932, that he was responsible for the idea because of papers he published during his *annus mirabilis* in 1905, Einstein responded that "Yes, I may have started it, but I regarded these ideas as temporary. I never thought that others would take them so much more seriously than I did." Later, Einstein put it this way to James Franck: "I can, if the worse comes to the worst, still realize that the Good Lord may have created a world in which there are no natural laws. In short, a chaos. But that there should be statistical laws with definite solutions, i.e. laws which compel the Good Lord to throw the dice in each individual case, I find highly disagreeable."

Einstein was not alone in being uncomfortable with the statistical nature of quantum mechanics and since then the vast literature that has appeared on the foundations of quantum mechanics was driven at least in part by an attempt to come to terms with the unusual and counterintuitive features inherent in the subject.

There were many attempts in the past to find hidden variables to avoid the statistical interpretation of the wavefunction, and there was even an informal monthly set of briefs called the *Epistemological Letters* put out by the *Association F. Gonset* or *Institut de la Methode*, which was distributed to, and contained theoretical correspondence from, some 100 prominent people in the field. In particular there was much discussion of Bell's theorem,[1] which ruled out the possibility of hidden variables, by Bell and others. I mention this in particular since those studying this period may not be aware of the past existence of this "Symposium", and it would be a very valuable resource for those working in the history of this area. Those who nonetheless choose to pursue the issue of hidden variables will sooner or later come across a poem written on the subject by Abner Shimony for a conference in the early 1970s (a Google translation for the poem is included below):[2]

---

[1] J.S. Bell, "On the Einstein Podolsky Rosen Paradox," *Physics* **1** (1964), 195–200; for a review of the subject, see: J.S. Bell, "On the problem of hidden variables in quantum mechanics," *Rev. Mod. Phys.* **18** (1966), 447.

[2] B. d'Espagnat (ed.), *Foundations of Quantum Mechanics* (Proceedings of the International School of Physics "Enrico Fermi", course 49), Academic Press, New York, 1971, pp. 56–76.

Tout le monde cherche les variables cachées	Everyone is looking for the hidden variables
Hélas, avec quel insuccès!	Alas, with what failure!
Elles sont timides, elles sont petites,	They are shy, they are small,
De courte durée, toujours en fuite.	Short lived, always on the run.
Elles sont partout en déguisement,	They are everywhere in disguise,
Empruntant bien des vêtements	Borrowing many clothes
Aux particules élémentaires.	Elementary particles.
Pour décider ce qu'on doit faire,	To decide what to do
Une assemblée de quarante mille	An assembly of forty thousand
Savants se tient à Célesteville.	Savants is held in Célesteville.
«Haute énergie!» Rataxès crie,	"High energy!" Rataxès cries out,
«Pour pénétrer le dernier nid	"To penetrate the last nest
De créatures si décevantes.»	Such disappointing creatures."
«Hourrah! les rhinoceros chantent,	"Hurray! the rhinoceros are singing,
«Agrandissons les cyclotrons!»	"Let's make the cyclotrons bigger!"
Babar pourtant conseille: «Non,	Babar, however, advises: "No,
La nature ouvre sa richesse	Nature opens up its wealth
Non par force, mais par finesse.	Not by force, but by finesse.
On verra les variables cachées	We will see the hidden variables
Aux rayonnements polarisés.»	Polarized radiation."
«Il a raison», dit Gregory,	"He's right," says Gregory,
Et la vieille dame fièrement sourit.	And the old lady proudly smiles.
Toute l'assemblée acclame son plan	The whole assembly applauds his plan
Et autorise avec élan	And emphatically authorizes
Un projet international,	An international project,
Créant le centre mondial	Creating the world center
Des appareils ingénieux.	Ingenious devices.
Les techniciens méticuleux	Meticulous technicians
Olur et Hatchibombotar	Olur and Hatchibombotar
Sous la conduite de Babar,	Under the leadership of Babar,
Commencent la grande expérience.	Begin the great experiment.
Pour en connaître les conséquences	To know the consequences
Variables cachées, oui ou non	Hidden variables, yes or no
Lisez la prochaine livraison.	Read the next issue.

There was no next issue and hidden variables were soon ruled out as a possibility. Babar the elephant first appeared in a French children's book written in 1931 by Jean de Brunhoff. The tale was made up and told to their children by Brunhoff's wife Cécile. Célesteville was the capital of Babar's kingdom where Olur was a mechanic and Hatchibombotar a street cleaner. Lord Rataxès, a rhinoceros, is the monarch of Rhinoland in Babar's kingdom.

While fully consistent with the usual quantum mechanics, a different interpretation of the wavefunction will be offered here. The classical conception of a point particle is replaced with one consonant with the quantum world.

Chapter 1

The Photon: History of a Misrepresentation

The Schrödinger equation describes the wave nature of matter and Schrödinger's approach had its origin in the works of Louis de Broglie, which will be discussed shortly. The solutions to this differential equation describe the motion of a particle and solving it gives the wavefunction ψ associated with the particle. The value of ψ is a function of the location in space and time chosen to evaluate it and the square of its modulus is conventionally interpreted to be a probability density which, when integrated over a volume, gives the probability of a particle being found in the volume of integration.

It is very important to realize that the wave properties of particles described by the Schrödinger wavefunction have nothing to do with waves that carry energy such as electromagnetic, acoustic, or water waves.

Consider electromagnetic radiation. The experimental situation is that light, or any electromagnetic radiation, displays a particle nature in that Einstein's photoelectric effect shows that it is composed of "photons". And this is where confusion often begins. The term "photon" is often taken to mean that electromagnetic radiation is composed of individual *particles* called photons.

What is true is that the radiation is composed of discrete energy packets whose magnitude is determined by their frequency. Intense radiation has enormous numbers of these packets, while the minimum energy that can be radiated is a single packet of energy $E = h\nu$, where h is Planck's constant[1] and ν the frequency of the wave. In short, the photon is not a particle!

---

[1] In 1900 Max Planck introduced the idea that the emission and absorption of radiation by matter takes place in finite quanta of energy, while Einstein, in 1905, maintained that this was an inherent property of radiation itself.

What picture does this bring to mind, for example, for radio waves? Consider the simplest case of dipole radiation. The wave pattern is composed of a vast number of photons, which are in phase with each other and each having an energy $h\nu$. Because the individual photons are generated from electrons with slightly different energies, the radiation has a bandwidth. Some of the radiated photons (a very large number) interact with the electrons in an antenna, thereby producing a detectable current.

It was Einstein who introduced the idea of a photon in an attempt to deal with the wave-particle dilemma early in the history of quantum mechanics. To quote Leon Rosenfeld, Einstein made the qualitative suggestion "that the photons, or the light quanta as they were called then, were some kind of singularity, of concentration of energy and momentum inside a radiation field. The radiation field would so to speak guide the photons in such a way as to produce also the interference and diffraction phenomena ..." This confused interpretation of a photon as a particle continues to this day. For massive particles, Einstein's suggestion of guidance is also found in the de Broglie–Bohm interpretation of quantum mechanics.

The problems raised by the concept of the photon are beautifully described by M. Sachs. Henri Bacry quotes him in his book *Localizability and Space in Quantum Physics* in the *Lecture Notes in Physics* series:

> "A very old, yet unresolved problem in physics concerns the basic nature of light ... Still, logical dichotomy and mathematical inconsistency remain in the usual answers to the question: What, precisely, is light?" [And a few pages later he discusses the conceptual difficulties.] "... a single photon, which, by definition, has a precise energy, is described mathematically in terms of a plane wave — a function that has an equally weighted value at all points in space at any given time. With this description, then, one would have to say that the single photon is everywhere, rather than somewhere — although it can be annihilated somewhere by looking for it at that particular place! Along

with this spatial description of the single photon, it is specified to be continually traveling at the speed of light. To the (perhaps naïve) inquirer, the logical difficulty appears in trying to answer the question: if the photon is everywhere at the same time, and is traveling continually on its own at the speed of light, where is it going to?"

And with regard to where the photon is, one can do no better than to again quote Henri Bacry:

"The photon is not localizable! It is not exaggerate [sic] to say that almost every physicist knows this fact but does not care. A position operator is not an important object. The important operators in quantum physics are the energy, the linear and angular momenta. The spectroscopist, whatever is his field (particle, nuclear or atomic), is not concerned with position! The position operator is only for students and, more precisely, only for beginners in quantum mechanics ... and for people interested in the sex of the angels, this kind of people you find among mathematical physicists, even among the brightest ones as Schrödinger or Wigner ..."

One does not need to know the details of position operators to understand the point of this quote!

Newton and Wigner[2] and Pryce[3] have given thorough discussions of position operators. It is the spin that is responsible for the photon's nonlocalizability; if the photon had spin zero, it would be localizable. Newton–Wigner derive an expression for the position coordinate for arbitrary spin, but for spin ½, it agrees with Pryce who defines the center of mass in coordinates where the coordinates taken in pairs have vanishing Poisson brackets. In such a frame, the total momentum vanishes,

---

[2] T. D. Newton and E. P. Wigner, "Localized states for elementary systems," *Rev. Mod. Phys.* **21** (1949), 400–406.

[3] M. H. L. Price, "The mass-centre in the restricted theory of relativity and its connexion with the quantum theory of elementary particles," *Proc. Roy. Soc.* **195A** (1948), 62.

and the center of mass is at rest — a result that is frame dependent. Note that the center of mass of a single particle is the same as the position of the particle. Pryce concludes, "From the point of view of relativistic quantum mechanics the only 'position vector' that has much interest is the one which is relativistically covariant ... The fact that its components do not commute leads to an uncertainty in the simultaneous measurement of order $\hbar/mc$." Or, as put by Bacry, "either it is impossible to measure any coordinate, that is, there is no position operator, or the position operator has three non-commuting components." In particular, massive particles with spin can be localized to a minimal uncertainty in one frame of reference, but in another frame, it will not be localized — localized states are not transformed into localized states under Lorentz transformations.

It might be useful to note here that the radiation field pattern or wavefunction of a single "photon" has a transverse character and related spin that is responsible for its nonlocalizability. As discussed earlier, this was shown historically by Newton and Wigner. They also found that localized states do exist for both massive and massless particles if their spin is zero. For massive particles that have spin, while a state can be localized for one observer it is not necessarily localized for another. As mentioned earlier, localized states are not transformed into localized states under a Poincaré transformation.

Does the photon really have a wavefunction associated with it, like massive particles? Perhaps somewhat surprisingly the answer is that it does and that its wavefunction can be used to show that the photon also displays the phenomenon of zitterbewegung or "trembling motion". It can be shown[4] that the photon wavefunction is given by

$$i\hbar\partial_t\psi = \mp c\left(\vec{S}\cdot\vec{p}\right)\psi .$$

---

[4] See my book *An Introduction to the Standard Model of Particle Physics for the Non-Specialist* (World Scientific, New Jersey, 2018), p. 124. Also: Zhi Yong Wang, Cai-Dong Xiong, and Qi Qiu, "Photon wave function and zitterbewegung," *Phys. Rev.* A **80** (2009), 032118; arXiv:0905.3420 [quant-ph].

Here ψ is a three-component spinor whose components S_i are scalar functions. The Hamiltonian for the photon is then $\mp c(\vec{S} \cdot \vec{p})$. For positive energy, one chooses the positive sign, which corresponds to positive helicity. The S_i turn out to be pure imaginary so that sign reversal corresponds to complex conjugation.

That the photon, a familiar electromagnetic wave, perhaps surprisingly displays zitterbewegung means that the phenomenon can be disassociated from the conception of a point particle. For massive particles, it is generally thought to be due to interference between negative and positive frequency states as was originally proposed by Schrödinger.

Kobe[5] has shown that for a single photon, which satisfies a relativistic analog of the Schrödinger equation, a velocity operator can be defined that has the photon moving with a constant velocity c and exhibiting an oscillation orthogonal to the photon's momentum with an amplitude approximately equal to the classical wavelength. This is identified as the photon's zitterbewegung and the spin of the photon is the associated orbital angular momentum.

---

[5] D. H. Kobe, "Zitterbewegung of a photon," *Phys. Lett.* A **253** (1999), 7–11.

Chapter 2

The Concept of a Particle*

What is a particle? We all know that the concept of a particle comes from Democritus' idea of atoms. His conception, and what today we would call Brownian motion, was related by Lucretius to the origin of all motion in his poem *On the Nature of Things* (50 B.C.E.):

> *Whence Nature all creates, and multiplies*
> *And fosters all, and whither she resolves*
> *Each in the end when each is overthrown.*
> *This ultimate stock we have devised to name*
> *Procreant atoms, matter, seeds of things,*
> *Or primal bodies, as primal to the world.*

• • •

> *For thou wilt mark here many a speck, impelled*
> *By viewless blows, to change its little course,*
> *And beaten backwards to return again,*
> *Hither and thither in all directions round.*
> *Lo, all their shifting movement is of old,*
> *From the primeval atoms; for the same*
> *Primordial seeds of things first move of self,*
> *And then those bodies built of unions small*
> *And nearest, as it were, unto the powers*
> *Of the primeval atoms, are stirred up*

---

* A small portion of this chapter also appears in an appendix of my book *An Introduction to the Standard Model of Particle Physics* (World Scientific, NJ, 2018)

By impulse of those atoms' unseen blows,
And these thereafter goad the next in size;
Thus motion ascends from the primevals on,
And stage by stage emerges to our sense,
Until those objects also move which we
Can mark in sunbeams, though it not appears
What blows do urge them.

With a little license, Lucretius' "*Procreant atoms, matter, seeds of things, Or primal bodies, as primal to the world,*" formed the basis of physical thought until quite late into modern times. In the ancient world, however, while it was accepted there might be different kind of atoms, the number of types was small and sometimes related to geometrical shapes. The advent of modern chemistry and spectroscopy in the 19th century began the formation of the current understanding of the nature of atoms.

Science up until the beginning of the 20th century could be traced back to its ancient Greek origins beginning perhaps in the sixth century B.C. Its evolution became what is known as the classical view of the world, and in particular, of physics. It also forced physicists, in particular, to learn a great deal more mathematics. But the use of sophisticated mathematics to describe the world raises some fundamental issues.

Mathematics studies the relations between arbitrarily defined abstract entities restricted only by the requirement that the definitions not lead to a contradiction. Mathematics now plays an enormous role in describing the physical universe, in general and in theoretical physics, in particular. But care must be taken since what may be true in mathematics is not necessarily a true reflection of the physical universe. If one can identify some of these "abstract entities" with elements of reality, then the mathematical relationships between the mathematical entities may contribute to an understanding of the physical interactions between these elements of reality with which the abstract entities are identified.

Consider complex numbers. There is nothing in the physical world to directly suggest that complex numbers would be useful in understanding the real world. Yet, we find, for example, that a complex Hilbert space along with a Hermitian[1] scalar product can be used to formulate two basic concepts in quantum mechanics: quantum states and observables. The quantum states are vectors in this Hilbert space and the observables are self-adjoint operators on these vectors.

Here is some history: The use of Hilbert space dates back to von Neumann. Max Jammer[2] has given an axiomatized presentation of von Neumann's approach, which incorporates Born's probabilistic interpretation of the wavefunction. Von Neumann originally assumed that there was a one-to-one correspondence between observables and self-adjoint operators. This was later abandoned in 1952 when G. C. Wick, E. P. Wigner and A. S. Wightman discovered the existence of "super selection rules". These restrict the set of self-adjoint operators that correspond to physically realizable states. When this is the case, it implies that the relevant Hilbert space decomposes to a direct sum of orthogonal subspaces.

The point of all this is to emphasize that how one interprets or maps elements of mathematical structures — particularly quantum mechanical ones — into the real world is far from trivial. The nature of space, time, and matter, as they are now understood, is very different from that of the classical world and it is these differences that lead to the difficulty in interpreting quantum mechanics.

The beginning of the 20th century brought with it two great revolutions in physics both due to Albert Einstein. The first was special relativity to be followed later by general relativity or the theory of gravity; the second was quantum mechanics initiated by Einstein's discovery of the photoelectric effect. The attempts to reconcile quantum mechanics with concepts brought over from classical mechanics has led to an enormous literature on the foundations of quantum mechanics and much confusion especially among non-physicists and students of physics. As mentioned in the Preface, this is due to the historical approach to teaching the subject coupled with the understandable struggle to carry over the basic concepts of particle and wave from classical physics.

---

[1] This is also spelled Hermitean by some authors.

[2] M. Jammer, *The Philosophy of Quantum Mechanics* (John Wiley & Sons, New York, 1974).

Today, it is believed that the elementary building blocks of matter are leptons and quarks, all of which are called fermions and obey the Dirac equation for a particle of spin of ½. In addition, there is electromagnetic radiation carrying a spin of 1. Lucretius' understanding of atoms has been carried over into the modern conception of "particle" although the basic fermions are thought to be "structureless" or "point" particles. Nonetheless, there have been many attempts to construct "classical" models for the electron.

Examples of attempts to maintain the concept of a point particle are the de Broglie–Bohm interpretation of quantum mechanics and the work of David Hestenes.[3] But retaining the idea of a massive charged point particle requires that both mass and charge be renormalized, a process that has never rested comfortably with many physicists.

The greatest challenge to the ancient idea of a particle came from the work of Louis de Broglie, who introduced in 1924 the idea that each particle had associated with it an internal clock[4] of frequency $m_0 c^2/h$. From this idea he found his famous relation showing particles of matter were associated with a wave. He did not believe a particle like the electron was a point particle, but rather that the energy of an electron was spread out over all space with a strong concentration in a very small region: "L'électron est pour nous le type du morceau isolé d'énergie, celui que nous croyons, peut-être à tort, le mieux connaître; or, d'après les conceptions reçues, l'énergie de l'électron est répandue dans tout l'espace avec une très forte condensation dans une région de très petites dimensions dont les propriétés nous sont d'ailleurs fort mal connues."[5] [Here is a rather free translation: The electron, we believe, perhaps wrongly, is known to us as an isolated piece of energy; but the energy of the electron is generally conceived to be spread throughout all of space with a very strong condensation in a region of very small dimensions, the properties of which are besides very little known to us].

---

[3] D. Hestenes, "The zitterbewegung interpretation of quantum mechanics," *Found. Phys.* **20** (1990), 1213–1232; "Electron time, mass and zitter," available on-line; "Zitterbewegung in quantum mechanics—a research program," arXiv:0802.2728 [quant-ph] 2008.
[4] This "internal clock" is also built into the Dirac equation.
[5] L. de Broglie, *Recherches sur la Théorie des Quanta* (Masson & Cie, Paris, 1963).

The concept of a wave being associated with the motion of elementary particles was introduced by de Broglie in his 1924 publication *"Recherches sur la Théorie des Quanta."* The hypothesis that matter as well as light have a wave-particle duality, and that this is a general property of microscopic particles, originates with him. What we call the wavefunction was called by de Broglie an *"onde de phase"* or a *phase wave*. It is a consequence of the relation $E = h\nu$. He also makes it clear that this wave cannot transport energy: *"qu'il ne saurait être question d'une onde transportant de l'énenergie."*

Notice that de Broglie first says that the energy of the electron is diffused throughout space and in the second quote that it is not a question of a wave that transports energy. Both cannot be true since any localization of the electron by an interaction means that the wavefunction must collapse to the local region of space where the localization took place. This essentially occurs instantaneously so that if the energy was diffused throughout all space, collapsing the wavefunction means energy would have to propagate faster than the speed of light. This problem led to the vast literature associated with the "collapse of the wavefunction," which is required in some interpretations of the wavefunction and not in others such as the many-worlds or ensemble interpretations.

Historically the root of the difficulty is the concept of electron waves where one makes an analogy with electromagnetic waves and constructs electron wave packets.[6] There, one sets the group velocity $v_g = \partial\omega/\partial k$ to be the classical particle velocity, and the phase velocity is $v_p = \omega/k$. In a non-dispersive medium, such as a vacuum, the angular frequency is proportional to the wave number so that the group and phase velocities are the same and equal to c. If the medium is normally dispersive, a small increase in wavelength results in an increase in phase

---

[6] For physicists, the idea of a wave packet comes from electrodynamics where, by use of a Fourier integral, one can superimpose waves that are plane-wave solutions to the wave equation derived from the source-free Maxwell equations. The derivation does not depend on the waves being electromagnetic in nature and the wave packets formed also apply to de Broglie "matter waves." The use of the term "matter waves" is unfortunate since it gives the impression that the waves carry energy. A clear exposition of the subject is given in J. D. Jackson *Classical Electrodynamics* (John Wiley & Sons, New York, 1999), 3rd Ed. Sect. 7.8.

velocity and the energy propagation is approximately the group velocity, which remains less than the phase velocity. If the dispersion is anomalous, this is no longer true and group velocity can exceed phase velocity.

In essence, the concept of electron wave packets should be rejected. An alternative conception of the wavefunction and its role in the motion of an electron is given in Chapter 4 titled *Matter and its Motion*.

There is some historical support for not relying on wave packets in the literature. In 1929 Mott[7] used the example of α-decay to show how the path in a Wilson cloud chamber due to α-decay need not involve the introduction of wave packets to explain the tracks observed. The problem is that the α-particle is represented as "a spherical wave which slowly leaks out of the nucleus." So how is a straight track produced by an expanding spherical wave? Intuitively, one would expect the wave to ionize gas atoms at random locations in the cloud chamber. His answer was that one must "consider the α-particle and the gas together as one system." He does this by defining the wavefunctions not in three-dimensional space but rather the multidimensional space formed by the coordinates of the α-particle and those of the atoms making up the gas. Mott does the calculations for two atoms, enough to establish the direction of the track. In the process, one need not consider the α-ray to be a particle at all. It is important to note that the direction of the track cannot be determined.

Let us return to the issue of the de Broglie phase wave. Its wavelength is given by the formula $\lambda = h/p$, where λ is the wavelength, p is the particle's momentum (mass times its velocity) and h is again Planck's constant. Notice that for $p = 0$, the wavelength is infinite, which implies that there is no oscillation and thus no phase wave.[8] What this tells us is that de Broglie's phase wave is related to a particle's motion through space and time. Wavefunctions describe how particles can travel through

---

[7] N. F. Mott, "The wave mechanics of α-ray tracks," *Proc. Roy. Soc.* **A126** (1929), 79–84; an analysis of this paper including a proof of Mott's result has been given by R. Figari and A. Teta in a SpringerBriefs in Physics volume titled *Quantum Dynamics of a Particle in a Tracking Chamber* (Springer, 2014).

[8] The discussion here excludes relativistic effects. A relativistic formulation would show that when a particle is stationary, it has a frequency of oscillation associated with it called the *zitterbewegung*, which de Broglie thought of as the inherent frequency of the electron.

space from one moment to the next and this motion is not deterministic, as it is in classical physics.

The connection of the phase wave with motion can also be seen by keeping in mind that since material particles have mass, special relativity tells us that we can always choose a frame of reference where the particle is at rest; i.e., we can catch up with a moving massive particle so that it is at rest with respect to us. This means that in one frame of reference, the particle has an associated phase wave while in another it does not. This is not the case for a wave carrying energy like electromagnetic radiation. There the velocity of propagation is the velocity of light and special relativity tells us that we cannot catch up with the wave and make it stop.

The historical gyrations on the meaning of the Schrödinger wavefunction are derived from the experimental fact that the quantum world, as captured in the wavefunction or other equivalent formulations, cannot be explained in terms of the classical concepts of a particle or wave. In trying to understand the meaning of the wavefunction, the first question that should be asked is whether it represents a single system or an ensemble of systems; i.e., does the wavefunction apply to the motion of a single particle or does it represent the relative frequencies resulting from measuring an ensemble of identically prepared systems. If one holds that the first is true, then there is the question of whether the wavefunction is a complete description of the system, raising the possibility that there may be unknown or hidden variables that could be specified to make the results consistent with the classical world. By now, as mentioned in the Preface, it has been established both theoretically and experimentally that the possibility of hidden variables can be ruled out by Bell's theorem. Bell's theorem basically deals with the concept of what is now known as entanglement, where the state of two quantum particles is correlated.

The second possibility, suggested and supported by Einstein, is that Born's statistical postulate should be accepted but interpreted so that the wavefunction applies to an ensemble of systems — an idea that others further developed. Louis de Broglie also introduced another idea where the wavefunction could be considered as a kind of "pilot wave" that guides an essentially classical particle into regions where the wavefunction is large. This concept was further developed by David Bohm, culminating in his by now classic papers that appeared in 1952. However, the de Broglie–Bohm theory has never been fully accepted by the scientific community.

Ultimately, we must accept the fact that an "elementary particle" is not a "particle" in the sense of classical physics; rather it is some form of spacetime excitation that can be localized through interactions, and yet — even when not localized, inherently obeys all the relevant conservation rules and *implicitly* retains "particle" properties such as mass, spin, and charge. This conception is a radical departure from the classical physics notion of a particle, which itself derives from our everyday perceptions and experience.

In what follows I will continue to use the word "particle" rather than make up a new word for the spacetime excitation corresponding to the "particle" for reasons of brevity, but this should be understood to be "particle" in quotes.

Even the name "elementary particle" is deceptive; perhaps "elementary excitation", or some such phrase, would pedagogically lead to less confusion. Instead, one is introduced to the concept of the "wave-particle duality". The problem is due to the use of ordinary language in trying to describe the quantum world. Max Born in his 1957 book *Atomic Physics*, put it this way: "The ultimate origin of the difficulty lies in the fact (or philosophical principle) that we are compelled to use the words of common language when we wish to describe a phenomenon, not by logical or mathematical analysis, but by a picture appealing to the imagination. ... Every process can be interpreted either in terms of corpuscles or in terms of waves, but on the other hand it is beyond our power to produce proof that it is actually corpuscles or waves with which we are dealing, for we cannot simultaneously determine all the other properties which are distinctive of a corpuscle or of a wave, as the case may be." Born's use of the word "interpreted" should be taken to mean what can actually be measured in an experiment. The attempt to interpret quantum phenomena in terms of classical concepts should be eliminated in pedagogy and the dual nature of the excitations of spacetime that correspond to elementary particles be taught from the first introduction of atoms in elementary school and the "solar system" model of the atom be eliminated at all educational levels.

The concept of "spin" is also a carry-over from classical mechanics to quantum mechanics of the concept of angular momentum like that of a spinning top. But unlike classical mechanics where angular momentum can take continuous values, in quantum mechanics, angular momentum is quantized so that, for example, spin angular momentum (the intrinsic angular momentum of a particle) can only take half-integral values (that is, 0, $\frac{1}{2}$, 1, ..., where these values are in units of $h/2\pi$).

One should not think of spin as the rotation of an elementary particle. As put by Born, "... the idea of a rotating electron, extended in space, possesses merely heuristic value; we must be prepared, on following out these ideas, to encounter difficulties. (For instance, a point at the surface of the electron would have to move with a velocity greater than that of light, if such values as have been determined experimentally for angular momentum and magnetic moment are to agree with those calculated by the classical theory.)" The heuristic value may have existed in the past, but today it is associated with the historical approach to teaching quantum mechanics and may introduce more confusion than insight.

And, in addition, there is the Pauli exclusion principle: While any number of integral spin particles can occupy the same quantum state, only two half-integral spin particles can occupy the same state, and then only if their spin is opposed. Thus, only two electrons can occupy the same state in atoms; this, coupled with the indistinguishability of electrons, is responsible for the existence of atoms and the periodic table of the elements. Put another way, the quantum numbers of two or more particles with half-integral spin cannot be the same.

Think of a single atom. Its nucleus is localized by the continuous interactions of its constituent components mediated by what is known as the strong force, distinguishing it from electromagnetic and other forces. The electrons surrounding it are localized by their interactions with the nucleus and each other, but only partially, up to the appropriate quantum numbers that describe stable atomic states as a function of distance from the nucleus and total angular momentum and its possible projections along the direction of a magnetic field, if one is present. One cannot localize electrons to definite positions in their "orbits" — that being yet another classical concept that does not apply to atoms. Two electrons cannot have the same n, l, j, and m quantum numbers.[9]

In general, the motion of a subatomic particle through space should be thought of as a sequential series of localizations along the particle's path due to interactions. It is not possible to define a continuous path in the sense of classical mechanics, only a series of "snapshots." Between localizations due to interactions, an elementary particle does not have a specific location. This is not a matter of our ignorance; it is a fundamental property of quantum mechanics; again, an "elementary particle" is not a "particle" in the sense of classical physics. One should not think of the particle existing between localizations due to interactions — there is no "classical little ball" being carried along by the de Broglie phase wave! To reiterate again: A particle is a spacetime excitation that can only be localized through interactions and which is characterized by its measurable "particle" properties such as mass, spin, and charge. The real mystery here is the nature of spacetime itself that allows such excitations to exist and have the properties they do.

One might think that the relationship between the classical Poisson bracket and the quantum mechanical commutator might shed some additional light on the transition from the classical world to the quantum mechanical one. If so, it is not obvious.

It is generally thought that the relation between classical and quantum mechanics is characterized by "letting $\hbar$ go to zero." For example, a standard problem in textbooks is to show that

$$\lim_{\hbar \to 0} \frac{1}{i\hbar}[A, B] = \{A, B\},\tag{1}$$

---

[9] In an atom, an *individual* electron may be characterized by four quantum numbers: $n = 1, 2, \ldots$; $l = 0, 1, 2, \ldots n - 1$; $j = 1 - 1/2$, $1 + 1/2$; $m = -j$, $-j + 1$, $\ldots + j$. n is known as the principal quantum number and is related to the distance from the nucleus; l is the angular momentum around the nucleus (orbital angular momentum); and j is the total angular momentum of a single electron, which combines its orbital angular momentum with its spin angular momentum. The quantum number m exists if a magnetic field is present, and designates the possible projections of j in the direction of the field. The details of the quantum numbers are not important for what follows.

where $[A, B]$ is the commutator and $\{A, B\}$ the Poisson bracket. This relationship applies independent of mass.

Note that the left-hand side of this equation, the commutator, is an operator on a Hilbert space and the right-hand side is a function.[10] It holds for most operators provided the Poisson bracket is considered to be an operator. And while there are some caveats, it always holds in the classical limit. What this tells us about the connection between the quantum world and the classical one is very far from clear. Nonetheless, it is worth checking where this relation comes from.

There are at least two ways to go to the classical limit: the first is to let $\hbar \to 0$ as above; and the second is to go to the limit of large mass. For a large mass, this equation reduces to $[A, B] = 0$ except for the case where $A = q$ and $B = p$, in which case one gets $[q, p] = i\hbar$ since $\{q, p\} = 1$.

Before showing where Eq. (1) comes from, a little additional discussion might be worthwhile. In classical and quantum mechanics, geometrical transformations — either Galilean or special relativistic — do not change what we consider to be the intrinsic properties of a particle. What this means, of course, is that there is a group property associated with the particle. The group of particular interest for quantum mechanics is the Poincaré group. The standard model of particle physics has enlarged this group, but the idea that a particle is associated with its group properties — introduced by Wigner[11] over fifty years ago — remains unchanged.

What Wigner showed was that the physically relevant representations of the Poincaré group with $p_0 \geq 0$ are parameterized by $s = 0, 1/2, 1, 3/2...$ for $m^2 > 0$ and $s = 0, \pm 1/2, \pm 1, \pm 3/2,...$ for $m^2 > 0$, where m is the mass and s the spin.[12] Thus, each kind of elementary particle is associated with a unitary irreducible representation of the Poincaré group. In a real sense, the particle and the representation are identified. As put by Sternberg, "an elementary particle 'is' an irreducible unitary

---

[10] The momentum and position take the form of operators on the l.h.s. of this equation and coordinates in phase space on the r.h.s.

[11] E. P. Wigner, *Ann. Math.* **40** (1939), 149.

[12] S. Sternberg, *Group Theory and Physics* (Cambridge University Press, 1994), Sect. 3.9.

representation of the group, G of physics, where these representations are required to satisfy certain physically reasonable restrictions . . . ."

While the invariance of the intrinsic properties of a particle under the Poincaré group applies equally well in classical and quantum mechanics, irreducible representations are usually only associated with a particle in quantum mechanics since spin is not quantized in classical mechanics. But as pointed out by Bacry,[13] Wigner did not restrict his approach to elementary particles, but referred to *elementary systems*. The example of an elementary system given by Bacry is that of the spin-zero hydrogen atom in its ground state with mass somewhat less than the sum of the proton and electron masses. While the set of all states of the hydrogen atom forms a representation space for a reducible representation of the Poincaré group, the proton and electron comprising the system no longer have irreducible representations associated with them since these particles are interacting and therefore do not form an isolated system.

One lesson to be learned from the above example is that collections of elementary particles in a particular state, while they may continue to be associated with an irreducible representation of the appropriate group, may lose some group properties like spin that are purely quantum mechanical in nature. What remains when the purely quantum mechanical properties are lost is the mass of the aggregate system. Going in the direction of decreasing mass, Rudolph Haag[14] has pointed out that "The physical interpretation of an irreducible representation of the Poincaré group (Newton and Wigner 1949) shows that the notion of a localized state of a particle becomes increasingly blurred with decreasing rest mass." Put the other way around, the localization of a particle is increasingly sharp as the mass increases. This can also be seen from the form of the Newton–Wigner position operators.

The original derivation of Eq. (1) was given by Dirac, but before turning to Dirac's derivation of this equation, consider the non-

---

[13] H. Bacry, *Localizability and Space in Quantum Physics*, Lecture Notes in Physics, No. 308 (Springer-Verlag, Berlin, 1988), Ch. 3; *Commun. Math. Phys.* **5** (1967), 97.

[14] Rudolph Haag, *Quantum Theory and the Division of the World, Mind and Matter* **2** (2004), 53; T. D. Newton and E. P. Wigner, *Rev. Mod. Phys.* **21** (1949), 400.

commuting matrices U, V, U_1, U_2, V_1, V_2. It is readily shown that the commutators $[U, V_1 V_2]$ and $[U_1 U_2, V]$ are

$$[U, V_1 V_2] = [U, V_1] V_2 + V_1 [U, V_2]$$
$$[U_1 U_2, V] = U_1 [U_2, V] + [U_1, V] U_2 .$$

(2)

Thus, the commutators on the left-hand side of these equations automatically satisfy the Leibniz rule. Dirac, in his derivation begins with Poisson brackets and when he arrives at the analog of the above, holds the order of the corresponding commuting dynamical variables fixed; i.e., having satisfied the Leibniz rule, he henceforth treats these variables *as if they were non-commuting matrices*. To be quite explicit, Dirac obtains the equations

$$\{U, V_1 V_2\} = \{U, V_1\} V_2 + V_1 \{U, V_2\}$$
$$\{U_1 U_2, V\} = U_1 \{U_2, V\} + \{U_1, V\} U_2 ,$$

(3)

and then *requires that the order of U_1 and U_2 be preserved in the second equation and the order of V_1 and V_2 in the first*. Dirac now evaluates $\{U_1 U_2, V_1 V_2\}$ in two ways using Eqs. (3), and subsequently equates the result to obtain

$$\{U_1, V_1\}[U_2, V_2] = [U_1, V_1]\{U_2, V_2\} .$$

(4)

Since U_1 and U_2 are independent of V_1 and V_2, Eq. (4) implies that

$$[U, V] = i\hbar\{U, V\} .$$

(5)

The value of the constant $\hbar$ is arbitrary and set by experiment and the factor i is introduced for the following reason: Dirac treats U and V as linear operators that could have an imaginary part and since the product of two real (i.e., Hermitian) operators is not necessarily real — unless they commute, Dirac introduces the factor of i to guarantee that $i(UV - VU)$ is real.

Instead of using Dirac's mixed approach of arbitrarily fixing the order of U_1 and U_2, and V_1 and V_2, as above, one can begin by initially treating these variables as non-commuting matrices in the Poisson bracket — some matrix representation of the invariance group. Treating U, V, U_1, U_2, V_1, V_2 as matrices results in

$$\{U, V_1 V_2\} = \{U, V_1\}V_2 + V_1\{U, V_2\} \text{ provided } \left[\frac{\partial U}{\partial q}, V_1\right] = \left[\frac{\partial U}{\partial p}, V_2\right] = 0$$

$$\{U_1 U_2, V\} = U_1\{U_2, V\} + \{U_1, V\}U_2 \text{ provided } \left[\frac{\partial V}{\partial q}, U_1\right] = \left[\frac{\partial V}{\partial p}, U_2\right] = 0.$$

(6)

These correspond to Dirac's equations given by Eqs. (3). Note that the vanishing of the commutators on the right-hand side of Eqs. (6) guarantees that the Poisson brackets on the left side obey the Leibniz rule. If $\{U_1 U_2, V_1 V_2\}$ is now evaluated *à la* Dirac, Eq. (5) is again obtained.

Thus, the requirements imposed by Dirac to derive Eq. (5) are equivalent to starting with non-commuting variables in the Poisson bracket to find a set of commutators whose vanishing guarantees that the Poisson brackets obey the Leibnitz rule. In the end, it is not clear how much light is shed by Eq. (1) on the transition from the classical world to the quantum mechanical one.

Chapter 3

Reinterpreting the Wavefunction

Problems with the conceptual foundations of quantum mechanics result, as discussed above, from the attempts by Niels Bohr and other physicists such as Werner Heisenberg and Max Born in the 1920s to continue to employ the classical concept of a particle in the context of the quantum mechanics, and in particular, in interpreting the wavefunction, given experimental observations.

Today, modern physics tells us that spacetime[1] supports a variety of excitations that can be identified with the various "particles" of matter whether short lived or stable; that the elementary building blocks of matter are leptons and quarks, which are fermions obeying the Dirac equation for a particle of spin ½. The proton and neutron are thus not true elementary particles.

The conceptual basis of the Dirac equation is directly related to the theory of groups. One can derive a classical semblance of the Dirac equation from the properties of groups and special relativity alone. This equation may be written as $(\gamma^0 p_0 + \gamma^i p_i - m)\psi(p) = 0$.

It corresponds to the relationship between the two spinors that come from the representations $(1/2, 0)$ and $(0, 1/2)$ of the Lorentz group.

The actual Dirac equation comes from the introduction of some minimal elements of quantum mechanics, namely $E = h\nu$ and $\lambda p = h$. If these relations are substituted into the expression for a classical wave packet, the resulting equations, obtained by taking separately a time derivative and the gradient, show that $E = i\hbar\partial_t$ and $\boldsymbol{p} = -i\hbar\boldsymbol{\nabla}$. Substituting these into the expression for the 4-momentum $p_\mu = (E, -\boldsymbol{p})$, turns the semblance of the Dirac equation into the actual Dirac equation.

---

[1] The nature of spacetime is, of course, crucial to the concept of an elementary particle or excitation. See Appendix A for a discussion of spacetime.

That the two equations $E = h\nu$ and $\lambda p = h$, needed to explain actual experiment, coupled with classical group theory give the Dirac equation shows what Eugene Wigner meant by "The Unreasonable Effectiveness of Mathematics in the Natural Sciences".[2]

For a free fermion, the Dirac wavefunction is the product of a plane wave and a Dirac spinor u, which is a function of the relativistic momentum p^μ, i.e., $\psi(x^\mu) = u(p^\mu)e^{-ip \cdot x}$. Substituting this wavefunction into the Dirac equation $(\gamma^\mu p_\mu - m)u(p) = 0$ for a particle at rest where $\vec{p} = 0$, results in the wavefunctions:

$$\psi^1 = e^{-imt}u^1, \psi^2 = e^{-imt}u^2, \psi^3 = e^{+imt}u^3, \psi^4 = e^{+imt}u^4,$$

where the eigenspinors u^i are given by

$$u^1 = \begin{pmatrix} 1 \\ 0 \\ 0 \\ 0 \end{pmatrix}, u^2 = \begin{pmatrix} 0 \\ 1 \\ 0 \\ 0 \end{pmatrix}, u^3 = \begin{pmatrix} 0 \\ 0 \\ 1 \\ 0 \end{pmatrix}, u^4 = \begin{pmatrix} 0 \\ 0 \\ 0 \\ 1 \end{pmatrix}.$$

As is readily seen, there are two different spin states for each of the energies $E = m$ and $E = -m$.[3] Take, for example, $\psi^1 = e^{-imt}u^1$, or explicitly,

$$\psi^1 = e^{-imt} \begin{pmatrix} 1 \\ 0 \\ 0 \\ 0 \end{pmatrix}.$$

This equation does not in any way mandate that one interpret it as a physical particle, only that the wavefunction has implicit in it the property of mass and a particular spin state. At this point, it is again worthwhile to emphasize de Broglie's description of the wavefunction as an "*onde de phase*" or a *phase wave*, and that "*qu'il ne saurait être question d'une onde transportant de l'énenergie*," that energy and, of course, mass are not carried by the wavefunction!

---

[2] E. P. Wigner, *Symmetries and Reflections* (Indiana University Press, Bloomington & London, 1967), p. 222.
[3] As per usual, the units are such that $\hbar = c = 1$.

The properties of mass and a particular spin state only play their role during an interaction with a second wavefunction. Rather than thinking of ψ as a function whose modulus squared is the probability density which, when integrated over a volume, gives the probability of a particle being in the volume of integration, one should instead interpret ψ as a function whose modulus squared is a probability density which, when integrated over a volume, gives the probability of an interaction taking place in the volume of integration. The real question, in the context of a physical elementary particle, is then about how the implicit mass and spin states embodied in a wavefunction become concrete during the interaction with a second wavefunction so that the usual rules, such as the conservation of mass and energy, are preserved.

And the answer is that it does not in the sense of a classical particle. *It is not necessary that the "collapse of the wavefunction" yield a classical particle in order for the interaction to occur.* The two wavefunctions alone interact in a way that the conservation rules are preserved. That is, the properties of each wavefunction (e.g., mass, charge and spin) combine to form a variety of new possible wavefunctions consistent with the conservation laws. Which combination actually exits the interaction region can vary. For example, when a high energy beam of particles from an accelerator interacts with a target, each interaction does not always result in the same set of particles.

We will see later that the key to understanding the interaction is the recognition of the fact that the phase of a wavefunction should be treated as a new physical degree of freedom dependent on spacetime position. If an electric field, for example, is present when a wavefunction representing a charged excitation of the vacuum comes under its influence, it affects the wavefunction by changing its phase as a function of position and hence the subsequent motion of the excitation.

In the rest frame of the excitation, the wavefunction has a spherical symmetry, while if it has momentum in some direction, the constant phase surfaces of the wavefunction will be elongated in the direction of the momentum. A series of localizations due to interactions with the electric field will show that the trajectory of the excitation follows that would be expected for a classical charged particle. *The wavefunction, at any given time, tells us the possible interaction locations in spacetime.*

One reason to revise the usual interpretation of the wavefunction is to avoid divergences. For point particles, quantum electrodynamics allows calculations of exceptional accuracy despite the divergences that occur in several types of Feynman diagrams, i.e., those dealing with radiative corrections where the diagrams have closed loops of virtual particles. An example is the photon self-energy diagram that is responsible for the phenomenon of vacuum polarization (also known as charge screening), which has no classical counterpart. The obvious question is how the revised interpretation of wavefunction given above would affect QED calculations.

When doing calculations, one generally works in the interaction "picture", also called the Dirac or Tomonaga picture. Using this picture, the Dyson operator (that relates the wavefunction at a time t_0 to that at time t_1) in the form of an integral equation is given by

$$\hat{U}(t, t_0) = 1 + (-i) \int_{t_0}^{t} dt' \hat{H}_1(t') \, \hat{U}(t', t_0),$$

where $\hat{H} = \hat{H}_0 + \hat{H}_1$ and $\hat{H}_0$ is for the free propagator.

Successive iteration of this equation gives a Neumann series, which one then time orders. The decomposition of the chronological product is then normally ordered so that all the creation operators appear to the left of all destruction operators. The $\hat{S}$ operator is then given by $\hat{S} = \hat{U}(-\infty, \infty)$. Normal ordering is equivalent to listing all the matrix elements of the scattering matrix S in a representation where the free-particle occupation numbers are diagonal. Feynman graphs are then a concise way of representing normal products.

The infinities encountered in QED are obtained by a process of renormalization. As an example, consider the electron self-energy. The full electron propagator $G_F(p)$ includes order-by-order the corrections to the Feynman Green function. The diagram for $G_F^1(p)$ looks like . The first step in renormalization is regularization, which makes the integral associated with $G_F^1(p)$ finite by introducing a parameter Λ, sometimes known as the "cut-off scale", which has the property that $G_F^1(p, \Lambda) \to G_F^1(p)$ as $\Lambda \to \infty$. $G_F^1(p)$ is then split into a divergent part and a finite part, which is known as a radiative correction and it is the finite part that leads to physically observable effects. The next step is the renormalization where the divergent part is incorporated into the tree-level propagator by a rescaling and a normalized mass parameter. One then takes the limit $\Lambda \to \infty$. In doing so, the original bare mass and charge parameters m_0 and e_0 become singular and the new physical parameters m and e are finite and the perturbation expansion becomes a series in e rather than e_0.

At this point, it is worth quoting Schweber, Bethe, and de Hoffman from Volume 1 of their 1955 book *Mesons and Fields*:

> "... even in the *absence* of infinities, we still would have to renormalize the theory. The origin of renormalization is due to the fact that we describe the state of the system in terms of unperturbed bare wavefunctions, whereas in the actual world we can never switch off the interaction between fields. Therefore, corrections to the bare mass and charge will occur. However, since only the bare mass (charge) *plus* the corrections to it can ever be observed, we must always express the observables in terms of the renormalized constants. In some sense, therefore, the questions of divergences and renormalization are separate ones. Nonetheless, since all local relativistic field theories with interactions are divergent, we shall use the term "renormalization" to express the fact that when the observable quantities are re-expressed in terms of the renormalized charge and mass, no divergences appear."

A second case of interest here is the diagram $\sim\!\!\bigcirc\!\!\sim$ for vacuum polarization or charge screening. This leads to the Uehling effect where the electrostatic potential of two charges in a vacuum is not precisely that given by Coulomb's law. The effect of vacuum polarization is to smear the effective charge of a point particle over a distance similar to the Compton wavelength h/mc. If one observes the charge at a distance large compared to this wavelength, the effective charge is less than e_0, while at distances increasingly smaller, the charge approaches the bare charge e_0. The Uehling effect also introduces a correction to the Lamb effect, the difference in energy between the $^2S_{1/2}$ and $^2P_{1/2}$ energy levels of the hydrogen atom, of -27MHz. Note that the Dirac equation predicts no difference between these states.

In the approach used to define the wavefunction introduced above point particles do not exist. Localizing the wavefunction by an interaction to less than the Compton wavelength means that the uncertainty in energy could exceed that of the rest mass associated with the vacuum excitation so that additional excitations could be created in the interaction. Thus, excitations could not be expected to be localized to much below the Compton wavelength. Nonetheless, the observable phenomena of QED, such as vacuum polarization, would remain undisturbed.

Rudolf Haag has thrown a monkey wrench into the edifice of quantum field theory with his eponymous theorem. It has to do with the interaction picture, which is used for calculations involving small perturbations of a well understood Hamiltonian in conventional Lagrangian field theory. Under the usual postulates and assumptions of quantum field theory, such as causality and the CPT theorem holding as well as the relation between spin and statistics, Haag's theorem essentially states that the interaction picture of canonical field theory cannot exist unless there are no interactions.[4] Nonetheless, the interaction picture is used to obtain the perturbation series often represented by the graphical notation introduced by Feynman. This approach to QED leads to reasonably accurate results in spite of Haag's theorem. But, except for those whose field is the Philosophy of Science, Haag's theorem seems to have been steadfastly ignored. After all, calculations are what matter!

What we have here is a perfectly reasonable set of axioms for QFT that are inconsistent in that they produce a result (the perturbation series) that correlates well with reality but which cannot exist because of Haag's theorem. This is reminiscent of the problems in mathematics resulting

---

[4] This, it should be noted, is not related to the problem of defining the S-matrix as the limit for $t_0 \to -\infty$, $t \to +\infty$ of $U(t, t_0)$.

from Gödel's incompleteness theorem,[5] which shows that there are statements whose truth or falsity cannot be proved in any axiomatic system strong enough to derive the natural numbers. More precisely, if the axioms are consistent, there are statements that are formally undecidable in the technical sense that neither the formula nor its negation can be formally deduced from the axioms.

Is there an analogy here? Could the real world exhibit effects related to Gödel's theorem? There is some evidence that undecidable problems do actually exist in physics as was shown by Cubitt *et al.*[6] The axioms of QFT are certainly consistent and one could speculate that in QFT, the "statement" would correspond to the perturbation series. Is this statement true or false? It is false given the axiomatic structure of QFT because of Haag's theorem, and true with regard to physical reality. Note that this has nothing to do with renormalization, which only multiplies the state vectors by constants and cannot change Haag's proof so as to introduce an interaction. This is, of course, not the same as is usually the case for Gödel's theorem where neither the statement's truth nor its falsity can be proved. Instead, we have a proof (Haag's theorem) of its falsity and a proof that it is true by comparing the calculated results to physical realty. Of course, only Haag's theorem follows from the axioms of QFT.

Interestingly enough, the well-known physicist Stephen Hawking *did* believe Gödel's theorem applies to physics. Here is a quote form a talk he gave in a public lecture at Texas A&M on 8 March 2017:

> "What is the relation between Gödel's theorem, and whether we can formulate the theory of the universe, in terms of a finite number of principles. ... Some people will be very disappointed if there is not an ultimate theory, that can be formulated as a finite number of principles. I used to belong to that camp, but I have changed my mind. I'm now

---

[5] K. Gödel, "Über formal unentscheidbare Sätze der Principia Mathematica und verwandter Systeme I" (On Formally Undecideable Propositions of Principia Mathematica and Related Systems I), *Monatshefte* (1931). A very readable book is: E. Nagel and J. R. Newman, *Gödel's Proof* (New York University Press, New York, 1960).

[6] T. Cubitt *et al.*, "Undecidability of the spectral gap," arXiv: 1502.04973v4 (15 June 2020); https://www.nature.com/articles/nature.2015.18983.

glad that our search for understanding will never come to an end, and that we will always have the challenge of new discovery. Without it, we would stagnate. Gödel's theorem ensured there would always be a job for mathematicians. I think M theory will do the same for physicists. I'm sure Dirac would have approved."[7]

How Wavefunctions Combine in an Interaction

The decay of parapositronium is a good example of how wavefunctions combine. Nonrelativistically, the wavefunction of positronium is the product of a spin vector and a Bohr atom wavefunction using the reduced mass for the positron and electron; relativistic corrections only differ by factors of two. The wavefunction for parapositronium, where the ground state has $n = 1$, $l = 0$, and a total spin of zero is given by $\psi_{n,l,m}(r)$ $\psi(S_{total}, s_z)$ where the spin part is given by

$$\psi(S_{total}, s_z) = \psi(0,0) = \frac{1}{\sqrt{2}}\left[e^+\left(+\tfrac{1}{2}\right) e^-\left(-\tfrac{1}{2}\right) - e^+\left(-\tfrac{1}{2}\right) e^-\left(+\tfrac{1}{2}\right)\right].$$

Notice here that one simply takes the product of the two wavefunctions to obtain a multiparticle wavefunction. In other words, the state space of the spin $1/2$ particle is given by the direct product of the particle's spatial state and the two-dimensional quantum space related to spin. The coordinates of the particle, which can only be localized to the $n = 1$ ground state, do not appear. The origin of the minus sign on the right-hand side of this equation will be discussed below.

Now photons, uncharged mesons, and uncharged bound states like positronium are eigenstates of the charge symmetry operator C. For the state of n photons, one has

$$C|n\,\gamma\rangle = (-1)^n|n\,\gamma\rangle.$$

Thus, parapositronium in the ground or singlet state, which decays with a C conserving process, can only decay into two photons.

---

[7] S. W. Hawking, *Gödel and the End of Physics*, http://yclept.ucdavis.edu/course/215c.S17/TEX/GodelAndEndOfPhysics.pdf.

We now turn to the origin of the minus sign in the spin part of the wavefunction for the ground state of parapositronium. Because the ground state has zero angular momentum, it has a single wavefunction, which could have either even or odd parity; scalar when the wavefunction does not change sign on a parity inversion, or pseudoscalar when it does change sign. Both parity and angular momentum are conserved when parapositronium decays.

The parity possibilities are distinguishable experimentally by the polarization of the two γ-rays produced by the annihilation of the electron and positron composing the parapositronium. The scalar possibility implies that the γ-rays are polarized in the same plane, while the pseudoscalar has the γ-rays polarized in perpendicular planes. Experimental results show that the polarization of the γ-rays is perpendicular so that the spin-zero ground state is a pseudoscalar, which then has odd parity. The electron and positron must consequently have opposite parity.

For a positron and electron pair, the charge symmetry operator is equivalent to charge exchange, which is the same as a parity (space inversion process) followed by a spin exchange. The order of the parity and spin inversion is irrelevant. Graphically, this looks as below:

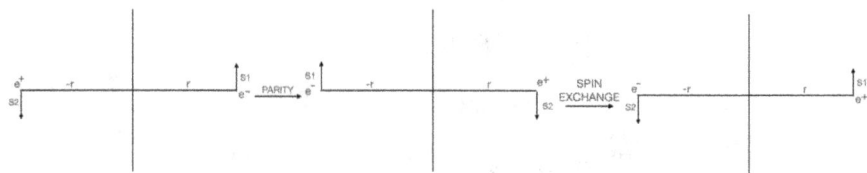

What the figure shows is that charge conjugation is equivalent to using the parity operator and spin exchange. This means that since charge is conserved in the singlet decay process, a singlet state is odd with respect to spin exchange.

Because the polarization of the two γ-rays is perpendicular, one can represent them as both being circularly polarized, thereby showing the direct connection with the spin of the electron and positron in the spin-zero ground state. As shown in the figure below, there are two possibilities.

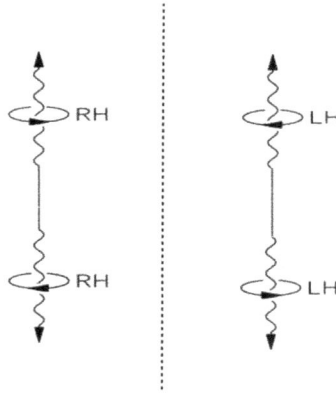

The two γ-rays from the annihilation are emitted in opposite directions so as to conserve momentum and must have the same handedness to conserve the spin-zero condition from the positronium ground state. The two possibilities differ from each other by a mirror inversion across the dotted line. Because the parity of the positronium is odd, the amplitude for the two possible decay processes must have opposite signs. If the γ-rays in the first decay process are labeled R_1 and R_2, and similarly for the second decay process L_1 and L_2, then the final state, F must be

$$|F\rangle = |R_1 R_2\rangle - |L_1 L_2\rangle$$

A parity inversion changes this state to

$$P|F\rangle = |L_1 L_2\rangle - |R_1 R_2\rangle = -|F\rangle.$$

Thus, the final state $|F\rangle$ has negative parity consistent with the initial spin-zero state of the positronium. The final state $|F\rangle$ should be compared with the ground state wavefunction $\psi(0,0)$ above.

Positronium was introduced here to help answer the question: How does the implicit mass and spin state contained in a wavefunction become concrete during the interaction with a second wavefunction so that the usual rules, such as the conservation of mass and energy, are preserved? As the discussion above shows, the concept of a classical point particle plays no role whatsoever in this example.

Again, the ground state of positronium has essentially the same wavefunction as a Bohr atom, if one uses the reduced mass for the positron and electron. In both cases, the wavefunction is generally considered to be a probability density for the two electrons in the case of a Bohr atom or of the positron and electron in the case of positronium. Both probability or density distributions have the geometry of a spherical shell. These are stationary states having no dipole moment and therefore do not emit electromagnetic radiation, which is not possible in Bohr's theory where the electrons revolve about the nucleus or the electron and positron revolve about each other in the case of positronium.

This contradiction was resolved in the case of the Bohr atom, by Max Born, as follows: "In wave mechanics this absence of emitted radiation is brought about by the fact that the elements of radiation, emitted on [sic] the classical theory by the individual moving elements of the electronic cloud, annul each other by interference." Presumably, "the elements of radiation" refer to the two electrons or possibly "the distribution of charge." There are two ways to interpret this: Born may have considered the absolute value of the wavefunction to be an actual "distribution of charge"; or that the "distribution of charge" corresponds to the time a particle spent in a given volume of space.

Whatever Born had in mind, this kind of interpretation led to calculations that gave superb results, but conceptually it is nonetheless the product of trying to retain the classical concept of a particle that has no real place in the quantum world.

Wavefunctions in Atoms

Once the Schrödinger equation for the hydrogen atom was solved, the solution for the helium would be expected to follow. This was and is not the case. So long as the electrons were considered to be classical point particles, albeit with quantum properties, the solution involved solving the three-body problem. And while very accurate approximations were developed, no exact solution has been found. It is the electrostatic interaction of the two electrons, involving the term $1/|r_2 - r_1|$, where r_1 and r_2 are the position vectors of the two electrons, that causes the difficulty. Without this term, the wavefunction would be separable; i.e., $\psi(r_1, r_2) = \psi_1(r_1)\psi_2(r_2)$. The solutions of the two wavefunctions are the same as that for the hydrogen atom with a nuclear charge of $2e$. If one assumes that both electrons are in their lowest energy states, these can be written as

$$\psi_0(r) = \frac{4}{\sqrt{2\pi}\, a_0^{3/2}} \exp\left(-\frac{2r}{a_0}\right),$$

where a_0 is the Bohr radius. The energy associated with each component of the separated wavefunction is $4E_0$, where $E_0 = -13.6\text{eV}$, the hydrogen ground state or ionization energy. This gives a value of -108.8eV for the ground state energy of helium, which is experimentally determined to be -78.98eV.

So, the repulsion between the electrons in the ground state of helium cannot be neglected. Physically, what is happening in the ground state is that each electron partially shields the nuclear charge from the other. Alternatively, one can view the repulsion of the electrons as contributing a positive potential energy partially offsetting the negative potential energy from the attractive force of the nuclear charge. In practice, one chooses a trial wavefunction like

$$\psi(r_1, r_2) = \frac{\mathbb{Z}^3}{\pi a_0^3} \exp\left(-\frac{\mathbb{Z}\,[r_1 + r_2]}{a_0}\right),$$

uses the variation principle, and then minimize the result with respect to $\mathbb{Z} < 2$, the effective nuclear charge number. By choosing more complicated trial wavefunctions with more adjustable parameters, one can get very close to the measured ground state energy.

Pragmatically, the effective shielding can be determined as follows: The energy needed to remove the first electron is 24.6eV; and for the second, it is 54.4eV, which is what one would expect by modeling the singly charged helium ion as a hydrogen atom with two protons in the nucleus. Since the hydrogen energy levels depend on the square of the nuclear charge, the energy needed to remove the remaining helium electron would be four times the ionization potential of hydrogen of

−13.6eV or −54.4eV, which is what is measured. The effective shielding can then be determined from $Z^2(-13.6\text{eV}) = -24.6\text{eV}$, or $Z = 1.34$.

Some of the difficulties described above in determining the helium ground state energy would not arise if the electrons were not considered to be point particles whose positions can be determined in their orbit. The electrons in helium are localized in space by their interactions with each other and the nucleus, but only up to the appropriate quantum numbers — n, l, m, and s, where for the parahelium ground state $n = 1$ and the rest of the quantum numbers are zero.

In any case, the Heisenberg uncertainty relation precludes localizing an electron to dimensions significantly smaller than the $n = 1$ Bohr radius of about half an Angstrom. At this radius the uncertainty in energy is already about 3eV. Further localization would soon exceed the ionization potential of hydrogen.

The spacetime excitation corresponding to an electron embeds in its wavefunction the properties of mass, charge, and spin, which become apparent during an interaction. The wavefunction itself, being a phase wave, does not itself carry these physical properties. In the case of the helium ground state, the only difference between the wavefunctions of the two electrons is the spin part where in the singlet state, the wavefunctions combine so that this spin part of the wavefunction, as was discussed above, is given by $\psi(S_{\text{total}}, s_z) = \psi(0, 0) = \frac{1}{\sqrt{2}}\left[e^+\left(+\frac{1}{2}\right)e^-\left(-\frac{1}{2}\right) - e^+\left(-\frac{1}{2}\right)e^-\left(+\frac{1}{2}\right)\right]$.

For two electrons the total wavefunction must be antisymmetric under an exchange of the particles, and since the singlet wavefunction must be symmetric, the spin part of the wavefunction must be antisymmetric as is shown in the above expression.

Rather than the individual point electron approach to the problem of atomic electrons, one can think of the electrons as standing waves with their charge distributed in space as a continuous distribution, proportional at any point to the squared magnitude of their wavefunctions. This is the approach used in modern quantum chemistry where one generally also uses the atomic orbital model to describe the electron charge distribution in matter. The orbitals of an electron in the hydrogen atom are shown below for different energy levels.

In the figure, the principal quantum number $n > 0$ is an integer that describes the primary energy level and could be associated with several orbitals, which make up what is often called the electron shells; l is the orbital angular momentum and is also a non-negative integer; and the magnetic quantum number m_l is an integer having the range $-l \leq m_l \leq l$ and describes the magnetic moment of the electron. In general, n determines the energy and size of an orbital, l determines the orbital's shape, and m_l its orientation. While the individual orbitals are generally shown as being independent of each other, the orbitals actually coexist at the same time.

It is interesting to note that Unsöld's theorem tells us that the sum of the electron densities for all orbitals of a given l of the same shell n, where all are occupied by an electron or an electron pair, the angular dependence vanishes and the density of the subshell with the same l has a spherical shape.

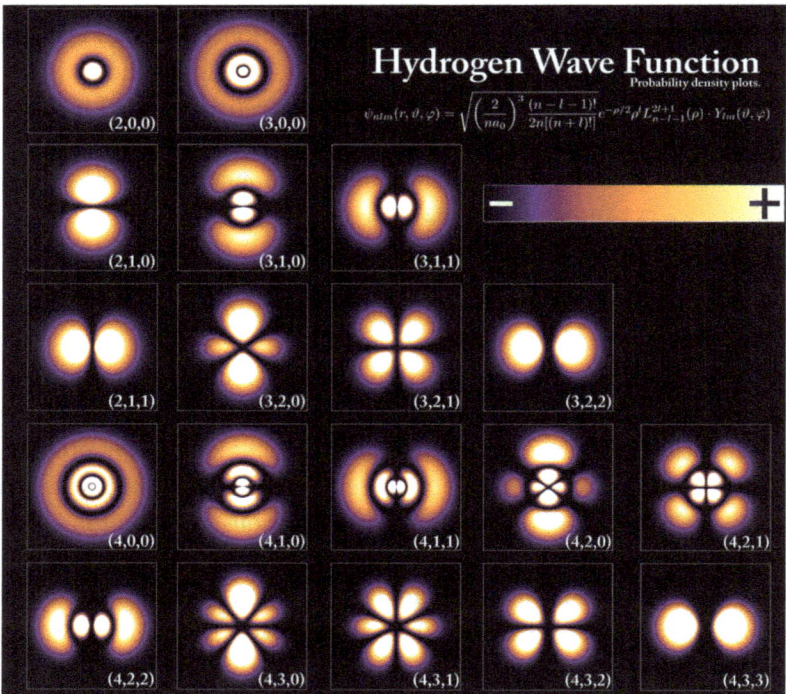

While the approach to interpreting the wavefunction as above may not be fully satisfying, it is certainly better than attempting to extend the concept of a classical particle into the quantum domain.

The examples in the above discussion were primarily about combining wavefunctions. In high energy collision of "particles," the collision may not be simple scattering where the final wavefunctions are the same as the initial ones associated with the incident "particles."

If the energy is high enough, new "particles" may be created that are consistent with the usual conservation laws. If the original wavefunctions are ψ_1 and ψ_2, the interaction would lead to $\psi_1 + \psi_2 \rightarrow \psi_3 + \psi_4 + \psi_5 \ldots$ In a scattering interaction, one would simply have $\psi_1 + \psi_2 \rightarrow \psi_1 + \psi_2$.

When the wavefunctions leaving the interaction region remain correlated, as in the Einstein, Podolsky, Rosen case, one can have a macroscopic quantum state. Currently, it is stated that the particles represented by these wavefunctions are "entangled." In the usual example, the orientation of the spin of two electrons prepared in a spin-zero state is not specified so that a measurement forcing one of the pairs into a specific orientation forces the other into the opposite orientation so as to preserve the spin-zero state of the pair. This happens instantaneously independent of the distance between the particles. Remember, the wavefunctions are phase waves *à la* de Broglie, whose velocities are not constrained by the speed of light. One also has macroscopic quantum phenomena in superconductivity and fluidity. For example, a macroscopic superconducting ring, which can carry a current for an indefinite period of time, is a quantum mechanical system that results from the occupation of a single quantum state and, when a current is present, the coherent motion of the Cooper pairs.

Wavefunctions in the Early Universe

It is generally believed that the universe expanded from a compact, dense, state where its energy was dominated by black body thermal radiation not matter. Because the photons interact only minimally, the radiation cannot reach thermal equilibrium. If the interaction with whatever matter there was is also negligible, the universe will expand homogeneously allowing the radiation to cool in an essentially adiabatic manner so as to preserve a thermal spectrum. This is because the heat capacity of the radiation[8] is far greater than that of the matter present.[9] This is the process thought to be responsible for today's isotropic 2.7 °K background radiation.

This is important since in the radiation dominated phase of the early universe, one conventionally *assumes* complete thermo-dynamic equilibrium[10] (which requires the presence of massive particles) and calculates the number density of the various particles produced thermally from the vacuum, where equal numbers of particles and antiparticles must be produced. Photons cannot alone produce pairs because of the need to conserve momentum and thus must interact with virtual particles from the vacuum to allow pair production. But, as discussed in Appendix B such vacuum fluctuations may not exist.

The fundamental particles resulting from the assumed fluctuations of the vacuum — quarks and leptons — are also thought to be point particles, and the radiation to arrive at their equilibrium is supposed to be electromagnetic and thus composed of photons. It should be remembered that photons are not particles and are not localizable. Photons would of course be localized should they interact with the quarks and leptons assumed to be present.

The evidence that the radiation is of a thermal nature comes from the expansion of the universe seen from the red shift of light from distant galaxies and the almost perfect 2.8 °K black body background radiation that dates from about 400,000 years after the birth of the universe. Previous to this time, the universe was opaque to all radiation.

---

[8] The specific heat of the radiation is $C_v = (16\sigma/c)T^3V$, where σ is the Stefan–Boltzmann constant.

[9] See P. J. E. Peebles, *Principles of Physical Cosmology* (Princeton University Press, New Jersey, 1993), Ch. 6. See also: R. E. Kelly, "Thermodynamics of black body radiation," *Am. J. Phys.* **49** (1981), 714–719.

[10] For a discussion and implications of assuming complete thermodynamic equilibrium in the Classical Hot Big Bang Picture, see: G. Börner, *The Early Universe* (Springer-Verlag, Berlin, 1993), Ch. 3.

This scenario, often called the "Big Bang" cosmological model, is a term that was coined by Fred Hoyle during a BBC radio program in 1949. He was thought by many to have used the term derogatorily.

When the expansion of the early universe, which lengthens the wavelength of all the photons, drops the temperature below a critical temperature, T_c, there is what is known as a phase transition where the actual existence of quarks and their subsequent binding into protons and neutrons becomes possible.

Note that if quarks had zero mass they would travel at the speed of light and their spin would be aligned either in the direction of motion or opposite to it. This chirality (handedness) is Lorentz invariant, and this symmetry is explicitly broken when the quark mass is not neglected. The critical temperature T_c corresponding to the chiral and confinement transitions (where quarks become bound) is thought to be similar.

Below T_c, the spacetime allows the existence of excitations that we identify with massive particles and their associated wavefunctions. Of course, the wavefunctions for photons exist both above and below the critical temperature.

Chapter 4

Matter and Its Motion

Quantum mechanically, motion consists of a series of localizations due to repeated interactions that, taken close to the limit of the continuum, yields a world line. If a force acts on a "particle", its probability distribution is accordingly modified. This must also be true for macroscopic objects, although now the description is far more complicated by the structure of matter and associated surface physics. Given the reinterpretation of the wavefunction in Chapter 3, how this will affect the issue on motion is the subject of this chapter.

The conceptual elements that comprise the presentation below are based on well founded and accepted physical principles, but the way they are put together — as well as the view of commonly accepted forces and the resulting motion of macroscopic objects that emerges — is unusual. What will be shown is that classical motion can be identified with collective quantum mechanical motion. Not very surprising, but the conception of motion that emerges is somewhat counterintuitive. After all, we all know that the term $\hbar/2m$ in the Schrödinger equation becomes ridiculously small for m corresponding to a macroscopic object.

Spacetime and Quantum Mechanics

What do we really know about the fundamental nature of spacetime? In essence, despite the enormous amount of material written about both, very little.[1] I have referred to spacetime rather than space and time as distinct entities because I will argue shortly that the two cannot be separated in any physically meaningful sense.

---

[1] See Appendix A.

To begin with, our concept of space itself is derived from the fundamental sense perceptions of the space around us, often formalized in science and mathematics as a manifold with a distance function that can be used to define a metric space. While conceptually very seductive, representing space in this way is also very deceptive in the sense that there is little reason to believe that the mathematical elements of points, neighborhoods, and open sets capture the essence of real physical space.

Space, without the concept of time, is — by definition — static. If space contains several objects, their orientation with respect to each other cannot change. To do so, the objects must move, and implicit in the concept of motion is time. Even if the change is instantaneous, there must be a "before" and "after". Think of any space-like hypersurface in spacetime. By definition — independent of what is contained in the hypersurface — it is static and changeless. Such a hypersurface cannot physically exist except in the limit where time slows to zero. Only null hypersurfaces are known to have this property. They do not, however, constitute a counter example because they could not exist outside the context of spacetime. While static space and an independent time exist as concepts, I maintain, but cannot offer a real proof, that in the real world, one could not exist without the other.

With the introduction of time, changes in the spatial location of objects become possible. Motion then depends on some type of wave equation, which means there is a wave front that propagates with finite velocity so that the region where the solution to the equation does not vanish may be localized in space and time.

For simplicity and clarity of exposition, the type of equations to be considered here will generally be linear equations of the second order.[2] These have linear differential operators of the form

$$L[\Psi] = a^{ik} \frac{\partial^2 \Psi}{\partial x^i \partial x^k} + b^i \frac{\partial \Psi}{\partial x^i} + C\Psi$$

---

[2] An unusually clear exposition is given by G. F. D. Duff, *Partial Differential Equations* (University of Toronto Press, Toronto, 1956).

where a^{ik}, b^i, and c are functions of x^i. If ξ_i is a covariant vector representing a surface element at some point P, from the a^{ik} one may form the invariant characteristic form relative to the operator $L[\Psi]$ given by $Q(\xi) = a^{ik}\xi_i\xi_k$. Under a coordinate transformation given by $\varphi(x^i)$, the surface element $\xi_i \equiv \partial_\varphi/\partial x^i$ transforms the quadratic form $Q(\xi)$ to

$$Q(\varphi) = a^{ik}\frac{\partial\varphi}{\partial x^i}\frac{\partial\varphi}{\partial x^k}.$$

Now the point of all this is that whether the equation is elliptic, hyperbolic, or parabolic depends respectively on whether the quadratic form $Q(\varphi)$: non-singular and positive definite, non-singular and indefinite, or singular in the sense that the determinant of a^{ik} vanishes. While the definite or indefinite character of the invariant quadratic form is independent of the coordinate system, because the coefficients a^{ik} may depend on the point P, an operator $L[\Psi]$ may change type at different points in space. However, if all of the a^{ik} are constant in some coordinate system, then the type (whether the operator is elliptic, hyperbolic, or parabolic) must be the same at all points. This is the case for many of the equations of importance in the physical sciences.

Since our interest is in wave equations, the a^{ik} of interest are those such that $Q(\xi)$, referred to its principal axes, contains only one sign differing from all the others. It is this one that is identified with time.

At this point, we have arrived at the pre-quantum mechanical space-time continuum of the early twentieth century. With the advent of quantum mechanics, the nature of the vacuum — usually identified with empty spacetime — dramatically changed. This occurred when it was found that at high energies or in strong fields, it was possible to create particles from the vacuum.

Today, the vacuum of spacetime is not considered to be empty, but is filled with various condensates. Among these is that composed of bound quark-antiquark pairs that goes under the name of "chiral symmetry breaking condensate." These are real particles not virtual particles associated with vacuum fluctuations. Data from the Supernova Cosmology Project has also shown that the vacuum contains a "dark energy" that differs little, if any, from Einstein's cosmological constant.

Quantum electrodynamics tells us that the vacuum is polarizable due to the presence of virtual electron-positron pairs and in quantum chromo-dynamics may be treated semiclassically by representing vacuum polarization by an effective dielectric constant, which gives the same result as quantum field theory. In essence, the modern concept of the vacuum has returned to that of some type of plenum, albeit one consistent with special and general relativity. In what follows, however, none of this will play a role. All that is required is spacetime as described above and non-relativistic quantum mechanics. Unfortunately, spacetime and quantum mechanics have consistency problems.

To deal with the concept of motion, we must begin with the well-known problem of inconsistency inherent in the melding of quantum mechanics and special relativity. One of the principal examples that can illustrate this incompatibility is the Minkowski diagram, where well-defined world lines are used to represent the paths of elementary particles while quantum mechanics disallows the existence of any such well-defined world lines. Despite this conceptual dissonance, the fusion of quantum mechanics and special relativity has proved to be enormously fruitful. This point has been made by Sklar[3] in his book *Space, Time, and Spacetime*: "Despite the rejection in quantum theory of the very notions used in the original justification of the construction of the spacetime of special relativity, it is still possible to formulate quantum theory in terms of the spacetime constructed in special relativity."

Feynman[4] in his famous paper "The Theory of Positrons" partially avoids the above conundrum, implicit in drawing spacetime diagrams, by observing that solutions to the Schrödinger and Dirac equations can be visualized as describing the scattering of a plane wave by a potential. In the case of the Dirac equation, the scattered waves may travel both forward and backward in time and may suffer further scattering by the same or other potentials. An identity is made between the negative energy components of the scattered wave and the waves traveling backward in time. This interpretation is valid for both virtual and real

---

[3] Lawrence Sklar, *Space, Time, and Spacetime* (University of California Press, Berkeley, 1974), p. 328.
[4] R. P. Feynman, *Phys. Rev.* **76** (1949), 749.

particles. While one generally does not indicate the waves, and instead draws world lines in Minkowski space between such scatterings, it is generally understood that the particle represented by these waves does not have a well-defined location in space or time between scatterings.[5]

The Feynman approach visualizes a non-localized plane wave impinging on a region of spacetime containing a potential, and the particle the wave represents being localized[6] to a finite region of Minkowski space by interaction with the potential. The waves representing the scattered particle subsequently spread through space and time until there is another interaction in the same potential region or in a different region also containing a potential, again localizing the particle. Even this picture

---

[5] This is best exemplified by the path integral formulation of non-relativistic quantum mechanics. The latter also has the virtue of explicitly displaying the non-local character of quantum mechanics.

[6] The use of the term "localization" is deliberate. There is no need to bring in the concept of measurements with its implicit assumption of the existence of an "observer." It is not necessary that an interaction having occurred, it needs to somehow enter human consciousness in order for the particle to be localized in space and time. The argument that it must enter human consciousness has been used, for example, by Kemble [E. C. Kemble, *The Fundamental Principles of Quantum Mechanics with Elementary Applications*, Dover Publications, Inc., 1958, p. 331] who states that "If the packet is to be reduced, the interaction must have produced knowledge in the brain of the observer. If the observer forgets the result of his observation, or loses his notebook, the packet is not reduced." It is not our purpose here to enter into a discussion of quantum measurement theory, but interpretations such as that expressed by Kemble often — but not always — rest on a lack of clarity as to what the wavefunction is assumed to represent. That is, whether the wavefunction applies to a single system or only to an ensemble of systems. While the ensemble interpretation has proven conceptually quite valuable in a number of expositions of measurement theory, it is difficult to understand how the wavefunction cannot apply to an individual system given the existence of many interference experiments using a series of *individual* electrons — where each electron participating in the production of the interference pattern must interfere with itself. Perhaps the most well-known attempt to bring consciousness into quantum mechanics is that of Eugene Wigner. The interested reader is referred to Wigner's book *Symmetries and Reflections* (Indiana University Press, Bloomington & London, 1967), Section III and references therein. In many circumstances a "measurement" done by an "observer" can be replaced by the role of the environment. The effect of such environmental decoherence has been discussed by Zurek and Halliwell: W. H. Zurek, "Decoherence and the transition from quantum to classical," *Physics Today* (October 1991); J. J. Halliwell, "How the quantum universe became classical," *Contemporary Physics* **46** (March–April 2005), 93.

is problematic since the waves are not observable between interactions. For the Dirac equation, the now famous Fig. 4.1 is intended to represent electron scattering from two different regions containing a scattering potential. The plane electron wave comes in from the lower left of the figure, and is scattered by the potential at A(3): (a) shows the scattered wave going both forward and backward in time; (b) and (c) show two second order processes where (b) shows a normal scattering forward in time and (c) the possibility of pair production. Feynman meant this figure to apply to a virtual process, but — as discussed by Feynman — with the appropriate interpretation, it applies to real pair production as well. Although the lines are drawn to represent these particles, no well-defined world lines exist.

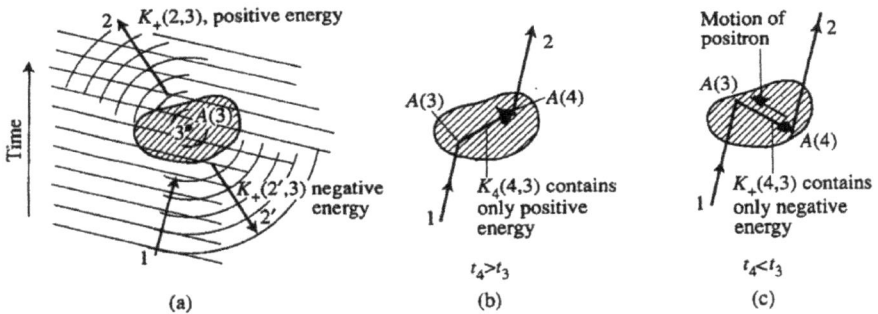

Fig. 4.1. Different electron scattering possibilities from a potential region. (a) a first order process, (b) and (c) second order. [Based on Fig. 2 of R. P. Feynman, "Theory of positrons," *Phys. Rev.* **76** (1949), 749–759.]

In a bubble chamber, for example, where the path followed by the charged particles is made visible by repeated localizing interactions with the medium, one would observe a pair creation event at A(4), an electron coming in from the lower left, and an annihilation event at A(3). Of course, since the particles involved here are massive, in the case of real pair production, the interval between A(3) and A(4) is time-like and the spatial distance between these events depends on the observer.

To reiterate, a world line is a classical concept that is only approx-imated in quantum mechanics by the kind of repeated interactions that make a path visible in a bubble chamber.[7] Minkowski space is the space of *events* — drawing a world line in a Minkowski diagram implicitly assumes such repeated interactions taken to the limit of the continuum.[8] While the characterization of Minkowski space as the space of events is often obscured by drawing world lines as representing the putative path of a particle in spacetime independent of its interactions, remembering that each point in Minkowski space is the position of a potential event removes much of the apparent incompatibility between quantum mechanics and special relativity, but it leaves us with a revised view of what constitutes motion.

Quantum Mechanical Motion

The picture of motion that emerges after the melding of quantum mechanics and special relativity is very unlike that of the classical picture of the path of a massive particle — like a marble — moving in spacetime. Consider a Minkowski diagram showing the world lines of

---

[7] Because the discussion to follow will give a different picture of a particle path, this is a good point to illustrate how motion is often described in quantum mechanics. Bohm [D. Bohm, *Quantum Theory*, Prentice-Hall, Inc., NJ, 1961, p. 137.] in describing how a particle path is produced in a cloud chamber maintains that ". . . when the electron wave packet enters the chamber, it is quickly broken up into independent packets with no definite phase relation between them . . . the electron exists in only *one* of these packets, and the wavefunction represents only the *probability* that any given packet is the correct one. Each of these packets can then serve as a possible starting point for a new trajectory, but each of these starting points must be considered as a separate and distinct possibility, which, if realized, excludes all others." If the particle has large momentum, ". . . the uncertainty in momentum introduced as a result of the interaction with the atom results in only a small deflection, so that the noninterfering packets all travel with almost the same speed and direction as that of the incident particle." [emphasis in the original]

[8] There is a considerable — and quite interesting — literature dealing with repeated "measurements" of a particle and what is known as "Turing's Paradox" or the "Quantum Zeno Effect." See, for example: B. Misra and E. C. G. Sudarshan, *J. Math. Phys.* **18** (1977), 756; D. Home and M. A. B. Whitaker, *Ann. of Phys.* **258** (1997), 237, lanl.arXiv.org, quant-ph/0401164.

several marbles at different locations. Given a space-like hypersurface corresponding to an instant of time in some frame, all the marbles would be visible at some set of locations. If one chooses a neighboring instant of time, these marbles would all still be visible at slightly different locations. This is because of the sharp localization of the marbles in space and time due to the continual interactions of their constituent components. Now consider the case of several elementary particles such as electrons. On any space-like hypersurface, the only particles "visible" would be those that were localized by an interaction to a region of spacetime that included the instant of time corresponding to the hyper-surface.[9] After any localization, the wavefunction of a particle spreads both in space and in either direction in time. Consequently, neighboring hypersurfaces (in the same reference frame) corresponding to slightly different times could have a different set of particles that were "visible." If motion consists of a sequential series of localizations along a particle's path, it is not possible to define a continuum of movement in the classical sense — there exists only a series of "snapshots."

Haag,[10] has put this somewhat in different terms: "The resulting ontological picture differs drastically from a classical one. It sketches a world, which is continuously evolving, where new facts are permanently emerging. Facts of the past determine only probabilities of future possibilities. While an individual event is considered as a real fact, the correlations between events due to quantum mechanical entanglement imply that an individual object can be regarded as real only insofar as it carries a causal link between two events. The object remains an element of potentiality as long as the target result has not become a completed fact."

---

[9] The term "visible" is put in quotes as a short-hand for the physical processes involved: the interaction of the particle needed to localize it on the space-like hypersurface and the detection of that interaction by the observer. It should also be emphasized that locali-zation is in both space and *time*. Just as localization in space to dimensions comparable to the Compton wavelength corresponds to an uncertainty in momentum of $\sim mc$, localization in time must be $\geq h/mc^2$ if the uncertainty in energy is to be less than or equal to the rest mass energy. For electrons, this corresponds to $\geq 10^{-20}$ second.

[10] Rudolph Haag, "Quantum theory and the division of the world," *Mind and Matter* **2** (2004), 53.

It is important to emphasize that between localizations due to inter-actions, an elementary particle does not have a specifiable location, although — because it is located with very high probability[11] somewhere within the future and past light cones associated with its most recent localization — it would contribute to the local mass-energy density. This is not a matter of our ignorance, it is a fundamental property of quantum mechanics; Bell's theorem tells us that there are no hidden variables that could specify a particle's position between localizations.

As an example of how localization works, consider a single atom. Its nucleus is localized by the continuous interactions of its constituent components. The electrons are localized due to interactions with the nucleus, but only up to the appropriate quantum numbers — n, l, m, and s. One cannot localize the electrons to positions in their "orbits."

Implicit in the discussion above is that an "elementary particle" is not a "particle" in the sense of classical physics. The advent of quantum mechanics mandated that the classical notion of a particle be given up.

Above, the flat spacetime of special relativity was used in the discussion. When the spacetime curvature due to gravitation is included, Minkowski diagrams become almost impossible to draw: Given a space-like hypersurface, the rate of clocks at any point on the hypersurface depends on the local mass-energy density and on local charge. Compared to a clock in empty spacetime, a clock near a concentration of mass-energy will run slower and will run faster near an electric charge of either sign. Thus the hypersurface does not remain "planar" as it evolves in time. To draw world lines one must take into account the general relativistic metric. This is why one uses light cone indicators at points contained in regions of interest.

The concepts of quantum mechanical localization and the resulting picture of motion are especially important in discussing many-particle problems and the transition to the classical world. In considering the penetration of a potential barrier, for example, one often restricts the problem to a single particle and calculates the probability that it will be

---

[11] If one uses only positive energy solutions of the Dirac equation to form a wave packet, the probability of finding a particle outside the light cone nowhere vanishes, although the propagator becomes very small for distances greater than the Compton wavelength $\hbar/mc$.

found on the far side of the barrier. For the many-particle case, say the surface barrier of a metal treated as a free-electron gas in a smeared positive background — an example that will be relevant later in this chapter — one would find that those electron wavefunctions that have been localized on the far side of the barrier will contribute to a real negative charge density. This charge density will interact with the smeared positive background.

Force, Fields, and Motion

Fields in classical physics are defined in terms of forces on either massive particles — in the case of Newtonian mechanics, or charges in the case of electromagnetism. General Relativity changed our way of thinking about the gravitational field by replacing the concept of a force field with the curvature of spacetime.

Starting with Einstein and Weyl,[12] there have been many attempts to geometrize electromagnetic forces. In all these attempts, charge — like mass in Newtonian mechanics — is treated as an irreducible element of electromagnetic theory that must be introduced *ab initio*. Its origin is not really a part of the theory. It does, nevertheless, have a unique spacetime signature.[13] Charge of either sign causes a *negative* curvature of space-time. The Einstein–Maxwell system of equations does not, however, allow different *geometric* representations for the electric fields due to positive and negative charges. This is a direct result of the fact that the sources of the Einstein–Maxwell system are embodied in the energy-momentum tensor, which depends only on the (non-gravitational) energy density. Charge, due to its geometrical effect on spacetime, always enters as Q^2 so that both positive and negative charges affect spacetime in the same way. Consequently, the electric field due to positive and negative charges cannot be identified with distinct changes in spacetime geometry. This also follows from the fact that, if we ignore the very small curvature

---

[12] L. O'Raifeartaigh, *The Dawning of Gauge Theory* (Princeton University Press, Princeton, New Jersey, 1997), pp. 24–37.
[13] G. E. Marsh, "Charge, geometry, and effective mass," *Found. Phys.* **38** (2008), 293–300.

of spacetime due to the energy density of the field, only charged particles are directly affected by the presence of an electromagnetic field. Thus, a full geometrization of charge does not appear to be possible within the framework of the Einstein–Maxwell equations.

The advent of modern gauge theory, incorporating the concepts of symmetry breaking and compensation fields, radically changed the understanding of fields. The electromagnetic interaction of charged particles in particular could be interpreted in terms of a local — as opposed to global — gauge theory within the framework of quantum mechanics. Interpreting the electromagnetic field as a local gauge field takes into account the existence of positive and negative charges and gives a good representation of the electromagnetic forces. It also gives us a concept of the electric field somewhat more enlightening than the classical one where the field is defined as the ratio of the force on test charge to the charge in the limit when it goes to zero.

The key concept for representing the electromagnetic force as a gauge field is the recognition that the phase of a particle's wavefunction must be treated as a new physical degree of freedom dependent on the particle's spacetime position. The four-dimensional vector potential plays the role of a connection relating the phase from point-to-point. Thus, the vector potential becomes the fundamental field for electromagnetism. The Aharonov and Bohm effect is generally cited to prove that this potential can produce observable effects, thereby confirming its reality.

The "gauge principle", as it is often called, is well illustrated by considering the non-relativistic Schrödinger equation in the context of electromagnetism.[14] It is also possible to give a relativistic version of the argument that appears below.

The Schrödinger equation for a free particle,

$$-\frac{\hbar^2}{2m}\nabla^2\Psi(\vec{x},t) = i\hbar\partial_t\Psi(\vec{x},t)\,, \qquad (1)$$

---

[14] I. J. R. Aitchison and A. J. G. Hey, *Gauge Theories in Particle Physics* (Institute of Physics Publishing, Bristol and Philadelphia, 1993), 2nd edition.

is not invariant under the *local* phase transformation

$$\Psi(\vec{x}, t) \longrightarrow \Psi'(\vec{x}, t) = e^{i\alpha(\vec{x}, t)}\Psi(\vec{x}, t). \tag{2}$$

To be invariant under such a transformation, the free particle Schrödinger equation must be modified so that it no longer represents a free particle, but rather one moving under the influence of a force. For the case of electromagnetism, the free particle Schrödinger equation must be replaced by

$$\left[\frac{1}{2m}(-i\hbar\nabla - q\vec{A})^2 + qV\right]\Psi(\vec{x}, t) = i\hbar\partial_t\Psi(\vec{x}, t), \tag{3}$$

where $\vec{A}$ and V transform according to

$$\begin{aligned}\vec{A} &\longrightarrow \vec{A}' = \vec{A} + q^{-1}\nabla\alpha(\vec{x}, t)\\V &\longrightarrow V' = V - q^{-1}\partial_t\alpha(\vec{x}, t)\end{aligned} \tag{4}$$

when $\Psi(\vec{x}, t) \longrightarrow \Psi'(\vec{x}, t)$.

The essence of the "gauge principle" is that demanding invariance under a local phase transformation corresponds to the introduction of a force. Of course, one can argue in the reverse: the introduction of a force can be represented as a local phase transformation. A simple example will be given below.

A free particle at rest samples a volume of space *at least* as large as its Compton wavelength, and the wavefunction associated with this sampling is such that a spherical volume is sampled in the absence of external forces. One might think here of a Gaussian packet (the lowest order wavefunction for the simple harmonic oscillator) which has the property of minimizing the uncertainty in both x and p, thereby giving the maximum localization possible.

If a force acts on the particle — say along the x-axis — this symmetry is broken by an extension of the probability distribution (the volume sampled) along the x-axis. To actually be "seen" to move, the particle must participate in a series of interactions so as to repeatedly localize it along its path of motion. If the force acting on the particle is modeled as a virtual exchange of quanta, such an exchange — viewed as an interaction — would serve to localize the particle. The propagation of a Gaussian wave packet representing the propagation of a charged particle under the influence of a constant force is an example well worth discussing further. This problem has recently been extensively treated by Robinett[15] and Vandegrift.[16]

The Gaussian wave packet $\psi_0(x, t)$ is a solution to the free-particle, one-dimensional Schrödinger equation

$$i\hbar \partial_t \Psi(x, t) = -\frac{\hbar^2}{2m} \frac{\partial^2 \Psi(x, t)}{\partial x^2} - Fx\Psi(x, t) \tag{5}$$

with $F = 0$. This solution has the property that it will remain centered at $x = 0$ for all values of t. Now let F be a time-independent, uniform force, implying a constant acceleration. In classical mechanics, such a force has the kinematic relation

$$x(t) = x_0 + v_0 t + \frac{1}{2} a t^2 , \tag{6}$$

where x_0 and v_0 are the initial position and velocity, and a is the acceleration. What Vandegrift shows is that the Gaussian packet solution to the Schrödinger equation with F being a uniform force becomes a wave packet centered at $x(t)$, that is

$$\Psi(x, t) = \psi_0 \left(x - x_0 - v_0 t - \frac{1}{2} a t^2, t \right) e^{iS(x,t)} \tag{7}$$

is a solution to the Schrödinger equation. The phase $e^{iS(x,t)}$ is a local phase transformation corresponding to $e^{i\alpha(x,t)}$ above, and $S(x, t)$ is explicitly given by

$$\frac{\hbar}{m} S(x, t) = v_0 x + axt - \frac{1}{2} a v_0 t^2 - \frac{1}{6} a^2 t^3 - \frac{1}{2} v_0^2 t . \tag{8}$$

[15] R. W. Robinett, "Quantum mechanical time-development operator for the uniformly accelerated particles," *Am. J. Phys.* **64** (1996), 803–808.

[16] G. Vandegrift, "Accelerating wave packet solution to Schrödinger's equation," *Am. J. Phys.* **68** (2000), 576–577.

This solution to Schrödinger's equation shows that the imposition of a uniform force is equivalent to making a non-relativistic transformation to an accelerating reference frame. It is also an example of the gauge principle.

Quantum Electrostatics

The gauge principle should also be able to explain macroscopic phenomena. The example to be used here will be that of electrostatics. Discussing electrostatics in a quantum mechanical framework is perhaps one of the most counterintuitive examples of collective quantum phenomena leading to classical behavior. What will be shown here is that the electric field, best interpreted as a phase field, affects the electron wavefunctions at the surface of a conductor and collectively this is what is responsible for the force acting on the conductor. Of course, one could simply use the classical electric field concept to achieve the same result, but — recalling the example of the Gaussian packet — greater insight into how the classical motion emerges is gained by considering the electric field as a phase field.

If a sphere holding a net positive charge Q is placed in an initially uniform electric field E_0, it will experience a force in the direction of the applied field. In solving Laplace's equation in terms of spherical harmonics, this force results from the term $Q/4\pi a^2$, where a is the radius of the sphere. The total charge density on the surface of the sphere is $3\varepsilon_0 E_0 \cos\theta + Q/4\pi a^2$. The electric field lines and associated surfaces of constant potential are shown in Fig. 4.2.

Notice that the constant potential surface corresponding to the potential of the sphere intersects the sphere and divides its surface so that those electric field lines terminating on negative surface charges are on one side of the intersection, and those whose origin is on positive surface charges are on the other. The value of θ giving the location of the intersection is given by the solution to the equation

$$\cos\theta + \frac{Q}{12\pi\varepsilon_0 E_0 a^2} = 0. \tag{9}$$

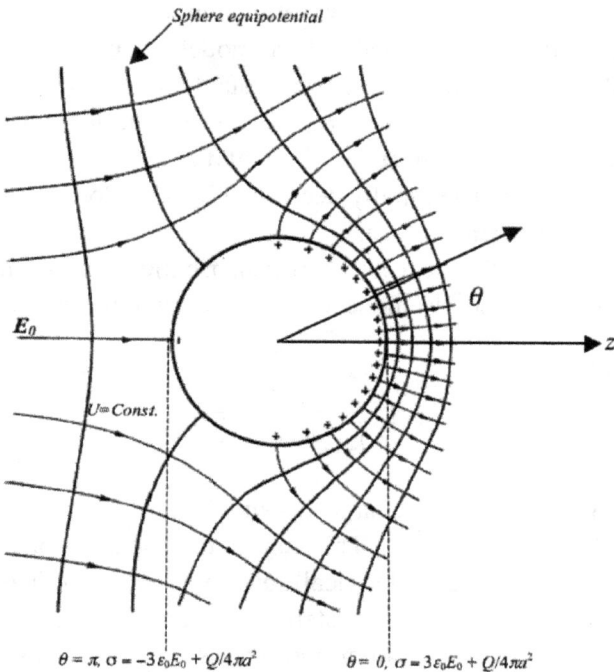

Fig. 4.2. Electric field E_0 and associated surfaces of constant potential U for a positively charged sphere of radius a in an initially uniform field. The induced surface charge density varies with θ, whereas that due to Q does not.

For $Q > 12\pi\varepsilon_0 E_0 a^2$, the electric field is directed outward from the surface for all θ. As can be readily seen from the figure, $\nabla U|_{\theta=0} > \nabla U|_{\theta=\pi}$, which implies a net force in the positive z-direction.

From a quantum mechanical perspective, the wavefunction at the surface is modified by the electric field interpreted as a phase field — similar to the example of the Gaussian wave packet discussed above. A net positive charge corresponds to removing a portion of the electron cloud of the nuclei near the surface, thereby unshielding these nuclei, which are the source of the positive charge. The wavefunction at the surface, as will be seen, is affected asymmetrically by the presence of an external electric field.

In order to calculate the wavefunction one has to simplify the problem and often the so-called jellium model[17] is used where the metal is modeled as a uniform positive background and an interacting electron gas. The surface of the metal is represented by the jellium (or geometrical) edge and is located at one half of the lattice spacing from the surface atom nuclei. The rapidly decaying electron cloud density extends beyond the geometrical surface.

The centroid of the excess charge distribution[18] (also known as the electrical surface) linearly induced by an external electric field is given by

$$z_{\text{ref}} = \frac{\int_{-\infty}^{\infty} x n_\sigma(x)\,dx}{\int_{-\infty}^{\infty} n_\sigma(x)\,dx}. \tag{10}$$

Here $n_\sigma(x)$ is the surface-charge density induced by the electric field perpendicular to the surface.[19] This centroid, calculated for $0 \le \theta \le 2\pi$, gives the position of the electrical surface — that surface where the external electric field appears to start. This surface is also the analog of the image plane (for the jellium and excess charge distribution) in the case of a plane conductor. Because we will be considering the position of the electrical surface at $\theta = 0$ and $\theta = \pi$, its position will be denoted by z_{ref}. The geometry and notional wavefunctions are shown in Fig. 4.3.

The symmetry of the electron-probability distribution along the x-axis is broken by the charge Q. This results in a change in the position of z_{ref}, which is determined by the net electric field due to the charge on the sphere and the external electric field.

---

[17] J. R. Smith, "Theory of electronic properties of surfaces," contained in: R. Gomer, ed., *Interactions on Metal Surfaces* (Springer-Verlag, New York, 1975). The term "jellium" was apparently first introduced by Conyers Herring.

[18] The excess charge distribution is also known as the screening or induced charge distribution.

[19] N. D. Lang and W. Kohn, "Theory of metal surfaces: Work function," *Phys. Rev.* B3 (1971), 1215–1223.

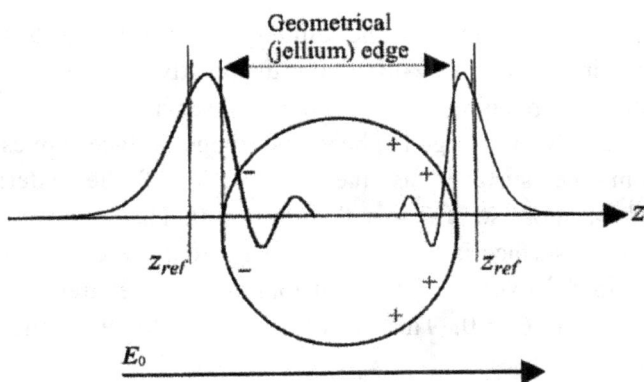

Fig. 4.3. Surface wavefunctions for a positively charged metal sphere in an *initially* uniform electric field E_0. z_{ref} is the centroid of the excess charge distribution at $\theta = 0$ and $\theta = \pi$. Notice that the location of z_{ref} is closer to the geometrical edge on positively charged portion of the sphere. The actual magnitude of the distance z_{ref} is around 3 a.u., or about 1.6 Å.

Thus, for an uncharged sphere placed in an initially uniform field, classically the gradient of the potential giving the field near the surface at $\theta = 0$ and $\theta = \pi$ is equal in magnitude, but opposite in direction with respect to the surface of the sphere. The net force therefore vanishes. Quantum mechanically, the location of the image surface is at the same distance from the jellium edge at $\theta = 0$ and $\theta = \pi$, so that the excess charge distribution interacts with the jellium equally yielding no net force in the direction of the field.

This will no longer be the case if the sphere is charged: the electric field at the surface due to the charge will asymmetrically sum up with that due to the external field E_0. For a positively charged sphere, the net field at $\theta = 0$ will be greater than at $\theta = \pi$. As the magnitude of the external electric field increases, the image surface moves inward towards the surface[20] at $\theta = 0$, but less so than at $\theta = \pi$. Because the electrical

[20] R. G. Forbes, "Charged surfaces, field adsorption, and appearance-energies: an unsolved challenge," *Journal de Physique* IV (Colloque C5, supplement au Journal de Physique III, Vol. 6 Septembre), (1996) C-25–C-30.

image surface is now closer to the jellium at $\theta = 0$ than at $\theta = \pi$, there is a net force in the positive z-direction (the positive charge on the sphere can be pictured as residing on the geometrical surface).

For a negatively charged sphere, the image surface moves outward (away from the surface) as the magnitude of the external field increases,[21] but more so at $\theta = \pi$ than at $\theta = 0$. The negative charge on the geometrical surface is then further away from the effective negative charge due to the excess charge distribution. The separation is greater at $\theta = \pi$ than at $\theta = 0$. This results in a net force in the negative z-direction.

Quantum mechanically, the origin of the force is similar to the example given earlier of the Gaussian packet, but in the case of the more complicated problem of a charged macroscopic sphere, one must adopt some simplifying model of the surface and its associated wavefunction. Above, the jellium model was used for the surface of a charged sphere with notional electron wavefunctions to illustrate the origin of the classical force. Thus, the collective force due to the asymmetric excess charge distribution that results from the localization of the underlying electron wavefunctions *is* the classical force.

Recapitulation

Quantum mechanically, motion consists of a series of localizations due to repeated interactions that, taken close to the limit of the continuum, yields a world-line. If a force acts on the particle, its probability distribution is accordingly modified. This must also be true for macroscopic objects, although now the description is far more complicated by the structure of matter and associated surface physics. The motion of macroscopic objects, as was illustrated in the context of electrostatic forces, is governed by the quantum mechanics of its constituent particles and their interactions with each other. The result may be characterized as such: collective quantum mechanical motion *is* classical motion.

---

[21] E. Hult *et al.*, "Density-functional calculation of van der Waals forces for free-electron-like surfaces," *Phys. Rev.* B **64**, 195414.

Since electromagnetic forces may be represented as a gauge field, electrostatic forces arise from the non-constant phase character of the electric field affecting many-particle wavefunctions. The example used was that of the force on a charged or uncharged metallic, conducting sphere placed in an initially constant and uniform electric field.

There is little that is new in this chapter. On the other hand, quantum mechanics is widely viewed as being imposed on the well-understood classical world of Newtonian mechanics and Maxwell's electromagnetism. This dichotomy is part of the pedagogy of physics and leads to much cognitive dissonance.

In the discussion above, the problems associated with transitioning from the reality of a quantum mechanical, many-body world to a classical one was avoided by using the jellium model — where a metal is modeled as a uniform positive background and an interacting electron gas. This was necessary since few if any many-particle problems with realistic interactions are exactly soluble. The theoretical approach to solving many-body problems often relies on the use of models to obtain approximations to specific problems. But care must be taken to determine the domain of validity of the model. Realistic interactions can create quantum correlations and collective states of matter — such as superfluidity and superconductivity — that have no counterparts in classical physics.

In the end, there is no classical world, only a many-particle quantum mechanical one that — due to localizations from environmental interactions — allows the emergence of the classical world of human perception. Newtonian mechanics and Maxwell's electromagnetism should be viewed as effective field theories for the "classical" world.

Appendix A

Spacetime

It was the mathematician Hermann Minkowski who joined space and time together in his 1908 talk to the 80th Assembly of German Natural Scientists and Physicians stating that "Henceforth, space by itself, and time by itself, are doomed to fade away into mere shadows, and only a kind of union of the two will preserve an independent reality." Interestingly enough, Einstein was not initially comfortable with the reformulation of special relativity by Minkowski, his former teacher, into four-dimensional spacetime. Let us begin with this union of space and time.

In Euclidean space, which has a positive definite metric, the time coordinate has the same status as the space coordinates; in relativity theory, the time coordinate has a special status due to the indefinite metric of Einstein spacetime. The most important thing to remember is that, just like the space coordinates, the time coordinate itself is not associated with a "flow" in any particular time direction. It does not have an intrinsic orientation, asymmetry, or arrow associated with it. Put another way, there is no "arrow of time" associated with the time coordinate itself except for what we give it for illustrative purposes.

The concept of "time" has multiple meanings: there is the coordinate itself; and there is the asymmetry of time in our three-dimensional space — which never changes its direction of flow; thermodynamic time, associated with the increase of entropy; psychological time, which each of us experiences as a present moment moving into the future; and finally, the concept of "cosmic time" associated with the expansion of the universe. Although these different concepts may be related, they are not identical and should not be confused.

The Minkowski diagrams of special relativity are made up of a continuum of space-like three-dimensional hypersurfaces along the time axis and perpendicular to it. The general view of time is that if one were to travel *backwards* in time one would see, for example, a sphere representing a propagating light pulse getting smaller and taking the size it had at an earlier time. That is, moving backward in time takes one to a three-dimensional space as it was in the past with the configuration of matter being what it was at each instant of past time. In this conception of time, three-dimensional hypersurfaces continue to exist in the sense that moving backward in time, were that possible, recapitulates three-dimensional space exactly as it was in the past. This concept of time leads to the usual conundrum that one could go back in time and murder one's grandfather. There is an even deeper problem.

The Einstein field equations of general relativity (the theory of gravity) have solutions that apply to objects like the earth or sun or to the universe as a whole. In the case of objects like the earth or the strong gravitational fields of massive neutron stars, these solutions have been tested to a very great accuracy. But the field equations also have perfectly good solutions, such as the infamous Gödel solution, that allow closed time curves.[1] Not only does this solution allow closed time curves, but in addition, closed timelike curves pass through every point of this space-time, and even more problematic is that there exists no embedded three-dimensional spaces without boundary that are spacelike everywhere, nor does a global Cauchy hypersurface exist.[2] Under the usual conception of time, moving in one direction along closed timelike curves is the equivalent to traveling backwards in time in that one may not only eventually arrive at the time when one began, but the configurations of three-dimensional space repeat themselves over and over again.

The Gödel solution and others like it are generally dismissed as being non-physical, but that simply begs the problem raised by their existence. The famous Kerr solution, representing the spacetime around a rotating mass, and which has no known interior solution — unlike the static Schwarzschild solution for a non-rotating mass — also has closed

---

[1] This solution caused enormous ferment in physics and philosophical circles. See: P. Yourgrau, *Gödel Meets Einstein: Time Travel in the Gödel Universe* (Open Court, Chicago, 1999); *A World Without Time: The Forgotten Legacy of Gödel and Einstein* (Basic Books, Cambridge, MA, 2005).

[2] A global Cauchy surface is a spacelike hypersurface such that every non-spacelike curve intersects it only once.

timelike curves if the angular momentum in appropriate units is greater than the mass, and one passes through the ring singularity. Yet, this solution is not dismissed as being non-physical.

Stephen Hawking has tried to get around the problem of closed time curves by introducing what he called the chronology protection conjecture: *"The laws of physics do not allow the appearance of closed timelike curves."* But thus far there has been no proof of this conjecture. Einstein's field equations alone, being partial differential equations, only tell us about the value of a function and its derivatives in an arbitrarily small neighborhood of a point. Whether closed time curves exist or not is a global question that may also depend on the topology of the spacetime. Some things about closed time lines are known. For example, for asymptotically flat spacetimes, if certain energy conditions are satisfied, closed timelike curves can only occur if spacetime singularities are present.

If a signal may be sent between two points in spacetime only if the points can be joined by a non-spacelike curve, then the signal is said to be causal (this type of formulation allows for the possibility that the two points can only be joined by light rays). The spacetime will be causal if there are no closed non-spacelike curves. The non-rotating solutions to Einstein's field equations, such as the Schwarzschild and Friedman–Robertson–Walker cosmological solutions are causally simple. For most "physically realistic" solutions it has been shown that the chronology condition — that there are no closed timelike curves — is equivalent to the causality condition stating that there are no closed non-spacelike curves.[3]

---

[3] While there is no need to discuss the time orientability of a spacetime here, it might be useful to give an example of a non-orientable spacetime. If one draws a circle representing the space axis along a Möbius strip and imposes a time direction perpendicular to the circle, after starting at any point and traversing the circle so as to return to the same point (on the other side of the strip of a paper model — a real Möbius strip only has one side), the time direction will be reversed. Such behavior implies that this $(1 + 1)$ dimensional spacetime is not time orientable. While a spacetime that is non-orientable has a double covering space, which is orientable, that does not eliminate the problem in the underlying base space. Covering spaces are very useful in mathematics, but in terms of the physics, it does not seem to be possible to jump between the two spaces. Either the base space or the covering space must be chosen as representing physical spacetime.

More generally, the Einstein field equations belong to a class of partial differential equations known as symmetric hyperbolic systems.[4] Such equations have an initial-value formulation in the sense that once initial data are specified on a spacelike hypersurface, the subsequent time evolution follows from this data. Unlike the Gödel solution, where a global Cauchy hypersurface does not exist, if a Cauchy surface does exist, and initial conditions are imposed on it for its future evolution governed by the Einstein field equations, closed timelines — the equivalent of a "time machine" — cannot occur. As put by Geroch, ". . . there exist solutions of Einstein's equation in general relativity that manifest closed causal curves. But we do *not*, in light of this circumstance, allow observers to build time-machines at their pleasure. Instead, we permit observers to construct initial conditions — and then we require that they live with the consequences of those conditions. It turns out that a 'time-machine' is never a consequence, in this sense, of the equations of general relativity, . . ."[5]

The usual conception of time, with its past three-dimensional hypersurfaces that continue to exist, imposes itself on our own psychological sense of time. But while we can remember and think of how things were in the past, this does not mean that the physical past continues to exist.

Another reason that the past, as conceived generally, does not exist has to do with microreversibility — the symmetry under time reversal — of the wavefunction given by Schrödinger's equation. For a system of particles, this symmetry is generally broken; i.e., the equation of motion describes the possible future evolution that a system may follow, but the time-reversal of the actual evolution of the system will not in general follow the same path backwards in time.

As pointed out earlier, just like spatial coordinates, the time coordinate itself is not associated with a "flow" in any particular time direction. If we measure the time distance around a closed timelike

---

[4] R. Geroch, "Partial differential equations of physics," arXiv:gr-qc/9602055v1 (27 Feb 1996).

[5] R. Geroch, *Advances in Lorentzian Geometry: Proceedings of the Lorentzian Geometry Conference in Berlin*, M. Plaue, A. Rendall, and M. Scherfner (Eds.) (American Mathematical Society, 2011), p. 59. arXive: gr-qc/1005.1614v1.

curve, there is no *prima facie* reason to expect the answer to be modulo the circumference.

Instead, one may think of the evolution of time as being a one-dimensional covering space over the original closed timelike curve as shown in Fig. A.1 below.[6] It is not necessary to identify the covering space as being the actual time curve in our universe since causality violations occur only if past three-dimensional hypersurfaces continue to exist. With this conception of time, one could go around a closed time curve many times without a causality violation. The need for Hawking's chronology protection conjecture is eliminated.

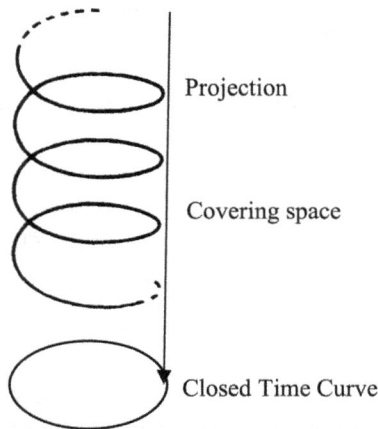

Fig. A.1. Closed time curve with a covering space. The original closed curve can be thought of as a projection of an *infinite* spiral over the closed time curve. Time changes monotonically in the covering space as one loops around the closed time curve.

Time and the Expansion of the Universe

The Friedmann–Lemaître spacetimes thought to represent our universe have exact spherical symmetry about every point, which implies that the

---

[6] A relevant concept is called "unwrapping": S. Slobodov, "Unwrapping closed timelike curves," *Found. Phys.* **38** (2008), 1082. Unfortunately, the process of extending a spacetime containing closed timelike curves generally introduces other pathologies.

spacetime is spatially homogeneous and isotropic, admitting a six-parameter group of isometries whose orbits are space-like three-surfaces (constant time) of constant curvature (positive, negative, or flat). One may choose the coordinates such that the line element has the form $ds^2 = dt^2 - R^2(t)dl^2$, where dl^2 is the line element of a time-independent Riemannian three-space of constant curvature, be it positive, negative, or flat, and $R(t)$ is the expansion function. What this form of the metric tells us is that the proper physical distance dl between a pair of comoving galaxies scales with time as $l(t) \propto R(t)$. For flat three-dimensional space, now believed to represent the actual universe, the function $R(t)$ monotonically increases with time.[7] One can readily show from the form of the metric that the velocity of separation of two comoving galaxies, V, is given by $V = \left[\frac{\dot{R}(t)}{R(t)}\right] l$, where the "dot" means the derivative with respect to time. This is the origin of the cosmological red shift. Thus, if $R(t)$ is constant, $V = 0$, and motion freezes.

The parameter t of the Friedmann–Lemaître spacetimes is explicitly identified with the time parameter used to express physical relationships such as in Newton's and Maxwell's equations. This implies that if the time is set equal to a constant number so that the universe freezes at some radius, the time associated with physical processes also freezes — nothing can propagate or change in three-dimensional space. Motion and the flow of time are inexplicably linked, as originally pointed out by Hermann Weyl. This would also be true in more general spacetimes where time may pass at different rates depending on the local mass-energy concentration. Thus, identification of the Friedmann–Lemaître time parameter (often called cosmic time) with physical time implies that the "flow" of time in three-dimensional space is due to the expansion of the universe.

The connection would seem to be even deeper. These spacetimes begin with an initial singularity — the term being used loosely. This means space and time came into being together and, in the real as opposed to mathematical world, may not be able to exist independently.

---

[7] The flatness of three-dimensional space does not necessarily imply that the full space-time is flat.

This is an obvious point: without time, there would be no space since expansion after the initial singularity would not be possible; and the expansion of space from the initial singularity implicitly introduces time and induces a time asymmetry in three-dimensional space.

This induced cosmic time is quite distinct from the thermodynamic time direction arising from increasing entropy when matter is present. It is also different from time as measured by "clocks" whose rate will vary according to both special and general relativity, but always in the implicit time direction induced in three-dimensional space by its expansion.

The Friedmann–Lemaître spacetimes can also have spatial sections that have positive curvature so that $R(t)$ is a cyclic function of time; i.e., the universe expands and then contracts (Fig. A.2). But time is not reversed during the contraction phase; the initial asymmetry in time persists.[8] Thus, either the expansion or contraction of the universe leads to a time asymmetry in the same direction. The term "expansion" alone will continue to be used here since the real universe appears to be flat.

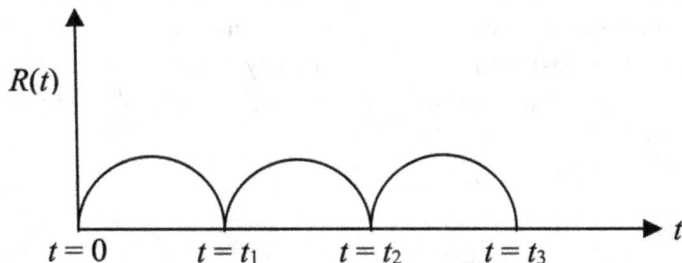

Fig. A.2. The Cyclic Friedmann–Lemaître spacetime with positive curvature that first expands from an initial singularity and then contracts to a singularity. The equation for $R(t)$ is a cycloid.

---

[8] In this connection, I should mention the work of Hawking who considered quantum gravity and metrics that are compact and without boundary. He showed that the observed asymmetry of time defined by the direction of entropy increase is related to the cosmological arrow of time defined by the expansion of the universe. S. W. Hawking, "Arrow of time in cosmology," *Phys. Rev.* D **32** (1985), 259.

If the singularities at the times indicated in Fig. A.2 are identified so that one has a closed time curve and the time asymmetry persists in the same direction as indicated in the figure, then time could increase monotonically through the cycles (see Fig. A.1 and associated discussion).

There is one additional point that should be made. The cosmological solutions to the Einstein field equations discussed above all have an initial singularity where spacetime itself is generally assumed to have come into being. The Einstein field equations themselves, however, do not inform us about what if anything existed "before" the initial singularity. The existence of the singularity simply indicates the limits of applicability of the field equations. In particular, these equations do not rule out the existence of some form of space or spacetime before the initial singularity. Most theoreticians assume that some form of quantum gravity will illuminate this issue. Unfortunately, current attempts in this direction — as exemplified by some forms of string theory, loop quantum gravity, non-commutative geometries, etc., have not had any convincing success. Also, to state a heretical view, there is no experimental evidence that space, time, or spacetime *is* quantized or that it need be quantized.[9] The desire to do so is primarily a matter of esthetics. It is based on the idea that because spacetime is a dynamical entity in its own right — due to its interactions with matter and energy — spacetime should in some sense be quantized.

The Asymmetry of Time

The standard "big bang" model of cosmology assumes that at the very beginning of the universe, there was no matter present but only energy in the form of enormously hot thermal radiation. The actual nature of this radiation, associated with a temperature similar to 10^{32} °K at the Planck time of 10^{-24} sec, is not really known, although it is generally characterized as thermal radiation, which is, of course, of electromagnetic

---

[9] One often hears Heisenberg's uncertainty relations incorrectly raised in this context. But they have to do with the theory of measurement in quantum mechanics and are directly derivable from classical wave theory and the relations $E = h\nu$ and $p\lambda = h$. They do not apply to the fundamental limits of spacetime itself.

origin. The extremely hot origin of the universe is confirmed by the existence of the isotropic 3 °K background radiation. The conversion of this early radiation into particle-antiparticle pairs, as the expanding universe cooled through a series of phase changes, is widely believed to be the source of the matter that exists today. The 3 °K background radiation itself comes from a time about half a million years after the initial singularity, by which time the plasma of ions (primarily hydrogen and helium, as well as electrons and photons) had formed and cooled to the point where it became a transparent gas. But there is a fundamental problem with this scenario that has not yet been resolved.

Consider the baryons (particles like neutrons and protons). From the observed ratio of the number of baryons to the number of photons in the background radiation — something like 10^{-9} — it is apparent that only a small fraction of the matter survived the annihilation of the particle-antiparticle pairs. This means that somehow there must have been a small excess of matter over antimatter before the annihilation occurred. For this to be the case, the symmetry between baryons and antibaryons must be broken. Baryon number conservation must be violated so that the various allowed decay schemes resulting in baryons can lead to a difference between the number of baryons and anti-baryons. The criteria for breaking this symmetry was established by Sakharov[10] quite some time ago: both C and CP invariance must be violated, or otherwise for each process that generates a baryon-antibaryon asymmetry there would be a C or CP conjugate process that would eliminate the possibility of a net asymmetry; and there must be a departure from thermal equilibrium, or CPT invariance — which must hold for any local, relativistic field theory — implying that there would be a balance between processes increasing and decreasing baryon number. There is some confusion in the literature about the meaning of the last requirement with regard to "time".

---

[10] A. D. Sakharov, *Pisma Zh. Eksp. Teor. Fiz.* **5** (1967), 32 [*JETP Lett.* **5** (1967), 24] [*Sov. Phys. Usp.* **34** (1991), 392] [*Usp. Fiz. Nauk* **161** (1991), 61]. Here C, P, and T are the discrete symmetries associated with charge, parity, and time respectively.

For example, Börner[11] states that, "Loosely speaking, the CPT-invariance of local, relativistic field theories and thermodynamic equilibrium imply the invariance under CP, because in thermodynamic equilibrium there is no arrow of time." Grotz and Klapdor[12] state that only if there is a departure from thermodynamic equilibrium will CP-violating interactions permit ". . . the rates of reactions which lead to the formation of baryons, to be larger than the rates of reactions which lead to antibaryons, but in thermodynamic equilibrium, no time direction is given, and the same would also apply to the inverse reactions."

Both statements argue that in thermodynamic equilibrium there is no Arrow of Time; i.e., no time direction is given. As it stands, this is certainly true, but as shown below in the discussion of thermodynamic time, this Arrow of Time has no relation to the kinematic time reversal transformation (see the book by Sachs referenced below). There is often confusion between the Arrow of Time and T-violation. As put by Sean Carroll in November 20, 2012 "blog" of the popular magazine *Discover*, referring to the recent results from BaBar on T violation, ". . . the entire phenomenon of T violation — **has absolutely nothing to do with that arrow of time** [emphasis in the original]."

With regard to Sakharov's requirement that there be a departure from thermodynamic equilibrium, Kolb and Turner[13] argue that, "The necessary non-equilibrium condition is provided by the expansion of the Universe. . . . if the expansion rate is faster than key particle interaction rates, departures from equilibrium can result." Calculations by Kolb and Turner show that only a very small C and CP violation can result in the necessary baryon-antibaryon asymmetry.

Systems in thermodynamic equilibrium (while they do not have an Arrow of Time) called "thermodynamic time" in this book, in the hopes of avoiding the kind of confusion found in the literature, do of course move through time in a direction given by the time asymmetry in the three-dimensional space within which we live.

---

[11] G. Börner, *The Early Universe* (Springer-Verlag, Berlin, 1993).

[12] K. Grotz and H. V. Klapdor, *The Weak Interaction in Nuclear, Particle and Astrophysics* (Adam Hilger, Bristol, 1990).

[13] E. W. Kolb and M. S. Turner, *The Early Universe* (Addison-Wesley, New York, 1990).

Because of *CPT* conservation, it is clear that *CP* violation means that *T*-invariance is also violated. Now these symmetry violations are generally discussed in the context of particle decays. For example, the decay of the *K*-meson tells us that the violation of *T*-symmetry is very small. But no matter how small the breaking of time reversal invariance, the fact that it exists at all implies that there is a direction of time in particle physics; i.e., a time asymmetry, which — to reiterate it once again — has nothing to do with the thermodynamic Arrow of Time.

Before beginning the discussion of the asymmetry of time in quantum mechanics, we turn to thermodynamic time so as to both complete the discussion above and introduce the Poincaré recurrence theorem.

Thermodynamic Time

Thermodynamic time has to do with the increase of entropy.[14] To begin with, the Poincaré recurrence theorem,[15] associated with thermodynamic and classical systems in general, states that for an isolated and bounded non-dissipative system, any particular state will be revisited arbitrarily closely; for macroscopic systems composed of many particles, the recurrence time will be very, very large. A simple example is a perfect gas confined to one side of a chamber by a membrane with the other side of the chamber being evacuated. If a hole in the membrane is opened, the gas will flow into the vacuum side; but ultimately all the gas will return to its original configuration after the Poincaré recurrence time has elapsed. From the point of view of thermodynamic time, it is possible to return to where the physical configuration of matter is *arbitrarily close* to its original configuration provided the assumptions given above on the nature of the system hold. What has been called cosmic time above always increases monotonically into the future even for such systems.

---

[14] An extensive and interesting discussion of time and entropy is contained in I. Prigogine, *From Being to Becoming: Time and Complexity in the Physical Sciences* (W. H. Freeman and Company, 1980).

[15] A clear and elegant proof of this theorem has been given by V. I. Arnold, *Mathematical Methods of Classical Mechanics* (Springer-Verlag, New York, 1989), p. 72.

Let us explore this issue more quantitatively. Consider a one-dimensional lattice of N particles of mass m elastically coupled to their nearest neighbors by springs with a force constant K, and let one particle have a mass $M \gg m$, which at $t = 0$ is given some velocity, the other particles being at rest. Rubin[16] computed the subsequent motion of the lattice and for large N found that the motion of the single particle with mass M was damped nearly exponentially. But the time symmetry is preserved and after a time $\frac{N}{2} \left(\frac{m}{K} \right)^{1/2}$ the lattice system completes a Poincaré cycle and returns to the original configuration at $t = 0$. A similar effect occurs with quantum systems as will be shown later in this book.

While a bounded system may therefore return to its initial state, there is no asymmetry in time involved. Nonetheless, one often hears of the "thermodynamic arrow of time" established by the second law of thermodynamics and the increase of entropy. The situation with thermodynamic time is quite murky. As put by Brown and Uffink,[17] "All traditional formulations of the Second Law presuppose the distinction between past and future (or 'earlier' and 'later', or 'initial' and 'final'). To which pre-thermodynamic arrow(s) of time were the founding fathers of thermodynamics implicitly referring? It is not clear whether this was a question they asked themselves, or whether, if pushed, they would not have fallen back on psychological time."

The idea that the thermodynamic arrow of time coincides with the psychological arrow of time led Hawking to observe that "... the second law of thermodynamics is really a tautology. Entropy increases with time, because we define the direction of time to be that in which entropy increases."[18] There has been some objection to this pithy characterization of the second law, but it suffices for our purposes. The connection between the thermodynamic arrow of time and the physics of time

[16] R. J. Rubin, *J. Amer. Chem. Soc.* **90** (1968), 3061.

[17] H. R. Brown and J. Uffink, *Stud. Hist. Phil. Mod. Phys.* **32** (2001), 525–538.

[18] S. Hawking, *The No Boundary Condition And The Arrow Of Time*, in J. J. Halliwell, J. Pérez-Mercador, and W. H. Zurek, (eds), *Physical Origins of Time Asymmetry* (Cambridge University Press, Cambridge, 1994).

reversal has been put quite succinctly by Sachs, "... the Arrow of Time has little to do with the time variable as measured by physicists. In particular, it has no bearing on the physics of time reversal."[19] The thermodynamic arrow of time will play no further role here. On the other hand, the Poincaré recurrence theorem will appear again in the next section.

Time Asymmetry in Quantum Mechanics

Below, in discussing Feynman's picture of the scattering of the Dirac wavefunction by a potential, waves will be allowed to travel backwards in time. These waves correspond to the negative energy states of the Dirac equation. That is, positrons may be interpreted as electrons propagating backwards in time. This may be explicitly shown by the transformation properties of the Dirac equation under the combination of parity, charge conjugation, and time reversal.

The Feynman interpretation of a positron as a backward-in-time moving electron is not inconsistent with the interpretation of time given above where past three-dimensional spacelike hypersurfaces do not continue to exist. The propagation into the past is very limited and the Feynman interpretation only applies to elementary particles. One way to accommodate this is to think of the three-dimensional space or hypersurface within which we live with a very small thickness in the time dimension.[20]

To simplify the discussion of time asymmetry in quantum mechanics, let us consider the Schrödinger equation $H|\Psi\rangle = i\partial_t|\Psi\rangle$. Like the Dirac equation, the probability amplitude Ψ is invariant under the T operator so that the physical content of the theory is unchanged. What will now be shown is that even though the physical content of quantum mechanics is preserved under time reversal (micro-reversibility under the

---

[19] R. G. Sachs, *op. cit.*
[20] R. P. Feynman, *Quantum Electrodynamics* (W. A. Benjamin, Inc., New York, 1962), pp. 84–85.

T operator), when one considers multiple systems, an asymmetry in time results. The discussion here follows that given by Davies.[21]

The Poincaré recurrence theorem associated with thermodynamic and classical systems that was discussed above has a quantum mechanical analog: Consider a collection of systems having only a ground state and one excited state whose energy can vary with the system. Now assume all systems are in their ground states save for one that is in its excited state. Assume further that all the systems are coupled by an interaction Hamiltonian H_{int}. After some time passes, there is a probability that the original excited system is in its ground state and one of the other systems is in its excited state. Davies finds that for two identical coupled systems, the Schrödinger equation gives a probability amplitude for the original excited system of $\cos^2(|H_{int}|t)$. Here the Poincaré recurrence period is $2\pi/|H_{int}|$. On the other hand, for a large number of systems, the probability amplitude for the original excited system is $e^{-2\pi|H_{int}|^2 t/\Delta E}$, where ΔE^{-1} is the density of states available. This is the usual time asymmetrical decay of an excited state with a half-life of $\Delta E/2\pi|H_{int}|^2$. As the number of systems increases, the density of states available goes to zero and the probability of the original state returning to its original excited state tends to zero.

While quantum mechanics satisfies what is known as the principle of micro-reversibility, processes that appear asymmetric in time are related to special initial conditions and the openness of the system, a good example being radioactive decay.

When we say we are "understanding" something, we generally mean we can relate it to something simpler that we already understand; and in the case of spacetime, this usually means quantum mechanics. And many attempts have been made to do this, none with outstanding success. All are based on the idea that general relativity tells us that spacetime is a dynamical entity, while quantum mechanics tells us that a dynamical entity has quanta associated with it, and consequently this entity can be in a superposition of quantum states. The implication is that there are "quanta" of space and time. But what does this mean? Does it mean that

---

[21] P. C. W. Davies, *The Physics of Time Asymmetry* (University of California Press, Berkeley, 1977), § 6.1.

space is made up of elemental little parcels of three-dimensional space? What role would time play with such parcels? Are there four-dimensional parcels of spacetime? Is time itself infinitely divisible? If not, is it made up of minimal steps? Is the ordering of such steps fixed?

The usual approach to quantum gravity is to treat the dynamical variable as being the spacetime metric $g_{ij}(x)$. Then the usual procedure of quantization leads to the infamous Wheeler–DeWitt equation, which DeWitt was known to refer to as "that damned equation". The Wheeler–DeWitt equation is essentially the Schrödinger equation for the gravitational field, and its wavefunction, $\Psi[g_{ij}(x)]$, is the "wavefunction of the universe". Time does not explicitly appear in the equation and there are conceptual problems with regard to the definition of probability, not to speak of the fact that the resulting theory is not renormalizable.

An analogy that may help with regard to these questions is to represent spacetime as a piece of cloth: from a distance, it is quite smooth, but as one comes closer, it begins to show the structure of its weave. The argument is made that if we look at space and time at the Planck distance and time, it would show a structure that we could understand and use to explain the nature of spacetime. It is string theory and loop quantum gravity that attempt to address these questions.

Some Metaphysical Thoughts

Notwithstanding the discussion above, there is little that is really known about the empty spacetime continuum itself — or the vacuum in the context of quantum field theory — except for what hints we have from special and general relativity, and those given by the Standard Model of particle physics. Unfortunately, the greatest fundamental conceptual issue with the Standard Model is that its redefinition of the vacuum begins to make it look like some form of æther, albeit a relativistic one! This results from the imposition of analogies from condensed matter physics, and in particular, superconductivity. Surely these analogies should not be taken literally. The fact that they "work" should only be taken as a hint about the real nature of the vacuum.

• • •

In the end, there could be limitations to the phenomenological approach of science to addressing epistemological or metaphysical issues. The situation with regard to our current understanding of space and time may, perhaps, be characterized by a portion of the ~1959 lecture of Professor Walter von der Vogelweide:[22]

> Introduction: "And now, ladies and gentlemen, Professor Walter von der Vogelweide will present *A Short Talk on The Universe*:
>
> Now, why, you will ask me, have I chosen to speak on the Universe rather than some other topic. Well, it's very simple. There isn't anything else!
>
> Now, in the universe we have time, space, motion, and thought. Now, you will ask me, what is this thing called time? [several second pause] *THAT* is time.
>
> Now, you will ask me, what is space? Now this over here — this is some space. However, this is not all space. However, when I said that was time that was all the time there was anywhere in the universe — at that time. Now, if you were to take all of the space that there is in the universe and *CRAM* it into this little tiny place, this would be *ALL* the space there was! Unless of course, some leaked out. Which it could. And did! Hence the universe!"

---

[22] From Severn Darden's *A Short Talk on the Universe*. This portion of professor von der Vogelweide's talk can be heard by clicking on http://www.gemarsh.com/wp-content/uploads/SpaceTimeM.mp3. The kind of improvisation that this slightly edited extract comes from began in the back of a bar called the *Compass* in Chicago's Hyde Park neighborhood near the University of Chicago campus. The *Compass Players*, including Mike Nichols, Elaine May, Shelley Berman, and Severn Darden, performed in Hyde Park from 1955–1958 and several of the members went on to form *The Second City Theater* in 1959.

Appendix B

Curved Spacetime: Illusory Particles?

Thus far it has been assumed that spacetime is the flat spacetime of Minkowski space. If a "particle" is real, one would expect it to exist near a massive body whether the spacetime is curved or not. In addition, one would expect that the coordinate frame used should not affect the existence of the particle nor should it generate particles. The Unruh effect and Hawking radiation, discussed below, show that this assumption could be incorrect.[1]

Vacuum Fluctuations

As was discussed earlier, both Schwinger and Pauli cast doubt on the reality of vacuum fluctuations. In Schwinger's source theory, the vacuum is "the state of zero energy, zero momentum, zero angular momentum, zero charge, zero whatever," and Pauli who stated that "it is quite impossible to decide whether the field fluctuations are already present in empty space or only created by the test bodies" and as late as 1946, he is quoted as saying that "zero-point energy has no physical reality."

There is some ambiguity about the term "vacuum fluctuations" and "zero-point energy" in the literature. If one is discussing the lowest-energy or ground state of some quantum mechanical system, the uncertainty principle tells us that the Hamiltonian must contain the term $\hbar\omega/2$.

If an electric, magnetic or vector potential field is present in the vacuum, the vacuum expectation of its field operator will vanish, but the

---

[1] A key reference for the Unruh and Hawking effects is: R. M. Wald, *Quantum Field Theory in Curved Spacetime and Black Hole Thermodynamics* (The University of Chicago Press, Chicago, 1994). Clear derivations are also given in: P. W. Milonni, *The Quantum Vacuum* (Academic Press, Inc., Boston, 1994).

expectation of the square of the field operators will not, which implies there are what are often called vacuum fluctuations of the field. In quantum field theory, each point in space has this zero-point energy associated with it, thus leading to infinite energy in any finite volume.

Zero-point energy, and fluctuations associated with it, may be eliminated by normal (Wick) ordering. But expectation values of normal-ordered operators vanish only for the *free* theory. In the interaction picture of quantum field theory, normal ordering eradicates Feynman diagrams with internal lines that begin and end on the same internal vertex. Higher order Feynman diagrams can be eliminated by what is known as complete normal ordering.[2]

It is often said that even the vacuum empty of all fields still retains the zero-point energy, whose average energy vanishes. What is left are the vacuum fluctuations of the so-called virtual particles that satisfy $\Delta E \Delta t \geq \hbar$ so that energy can be taken from the vacuum to allow particles to appear for very short times. These are the type of vacuum fluctuations that apply to the Unruh and Hawking effects and whose reality Schwinger and Pauli doubted.

More recently, Jaffe[3] has pointed out that the Casimir effect, often cited as the proof that vacuum fluctuations are real, and its experimental confirmation does not establish the reality of zero-point fluctuations. He points out that vacuum-to-vacuum Feynman graphs, that essentially define the zero-point energy, are not involved in the calculation of the Casimir force, which only involves graphs with external lines. In conclusion, he states that there is "no known phenomenon, including the Casimir effect, [that] demonstrates that zero point energies are 'real'."

Casimir calculated the attractive force between two uncharged parallel conducting plates due to the quantum electromagnetic zero-point energy of the normal modes between the plates. But whether the force

[2] J. Ellis, N. E. Mavromatos, and D. P. Skliros, "Complete normal ordering 1: Foundations," *Nucl. Phys. B* **909** (2016), 840–879.
[3] R. L. Jaffe, "Casimir effect and the quantum vacuum," *Phys. Rev. D* **72** (2005), 021301; arxiv:hep-th/0503158 (2005).

is attractive or repulsive depends on the geometry of the uncharged conductor.[4]

A discussion of how the vacuum is defined and the relation of vacuum fluctuations to the Casimir effect and the cosmological constant problem are contained in Appendix C.

The Unruh Effect

The Unruh or Davies–Unruh effect occurs for a uniformly accelerated detector in a vacuum that would measure a temperature given by $T = \hbar a/2\pi c k_B$, where a is the local acceleration and k_B is the Boltzmann constant. This temperature has the same form as the Hawking temperature of a black hole $T_H = \hbar g/2\pi c k_B$, where g is the surface gravity of the hole. In the case of the hole, there is spontaneous particle creation. Although the two effects are mathematically and physically distinct, they are superficially related by the equivalence principle between acceleration and gravitation.

The electromagnetic zero-point fluctuation of the vacuum may be regarded as a propagating electromagnetic field with a spectral energy density[5]

$$\rho(\omega)d\omega = \frac{\hbar\omega^3}{2\pi^2 c^3}d\omega .$$

There has been some debate over whether this field should be regarded as real or virtual. The evidence given to support the reality of the various contributions to the vacuum energy is the Casimir effect, which is a consequence of the lowest order vacuum fluctuations, and higher order effects like the Lamb shift. But there are alternative explanations. The Casimir effect could result from fluctuations associated with the constituents of the plates rather than vacuum fluctuations. Schwinger's source theory takes this point of view and avoids vacuum fluctuations in both the

---

[4] T. H. Boyer, *Phys. Rev.* **174** (1968), 1764. See also: W. Lukosz, *Physica* **56** (1971), 109; *Z. Physik* **258** (1973), 99.
[5] T. H. Boyer, *Phys. Rev.* **182**, 1374; P. W. Milonni, *The Quantum Vacuum* (Academic Press, Inc., Boston, 1994), p. 49.

Casimir and higher order QED effects. And Pauli's opinion was quoted above.

The spectral energy density given above is Lorentz invariant, but in a uniformly accelerated reference frame with proper acceleration a one finds a pseudo-Planckian spectrum with a radiation temperature $T = \hbar a / 2\pi c k_B$, the same as that for the Unruh effect. In this frame, the spectral energy density has the form

$$\rho(\omega)d\omega = \left[\frac{\omega^2}{\pi^2 c^3}\right]\left[1 + \left(\frac{a}{\omega c}\right)^2\right]\left[\frac{\hbar\omega}{2} + \frac{\hbar\omega}{\exp\left(\frac{2\pi c\omega}{a}\right) - 1}\right]d\omega \,.$$

Notice that at high frequencies this reduces to the previous expression. The Unruh effect is thus due to the lower frequencies.[6]

The two above spectral energy densities tell us that the vacuum state is not the same in an unaccelerated inertial frame and an accelerated one. But general relativity tells that the coordinates used are arbitrary and have no geometrical or physical meaning. As put by Misner, Thorne and Wheeler in their book *Gravitation*: "The laws of physics, written in component form, change on passage from flat spacetime to curved spacetime by a mere replacement of all commas by semicolons," where the comma represents ordinary partial differentiation and the semicolon covariant differentiation. The laws of classical physics are local in nature and the comma-goes-to-semicolon rule is directly related to the equivalence principle. If one believes vacuum fluctuations are real, then one must accept that not all coordinate systems are physically equivalent in quantum mechanics.

In spite of the enormous literature on the Unruh effect, this has been unpalatable for many and some maintain that the effect does not exist.[7] I will not go through the technical details of the argument, but simply give an outline.

---

[6] See B. Haisch, A. Rueda, and H. E. Puthoff, "Inertia as a zero-point-field Lorentz force," *Phys. Rev. A* **49** (1994), 678–694.

[7] See, for example, V. A. Belinskii *et al.*, *JETP Lett.* **65** (25 June 1997), 902.

The coordinates of a reference frame in Minkowski space accelerating with a constant proper acceleration are, in the context here, often called Rindler coordinates. The figure below shows the left (L) and right (R) Rindler wedges in Minkowski space. The hyperbolic paths of two objects undergoing constant *proper* acceleration are shown. The closer the hyperbola is to the origin, the greater the acceleration. The light cone represents the event horizons bordering the part of Minkowski space accessible to "Rindler observers". The interior of the R wedge, an incomplete manifold, is called Rindler space.

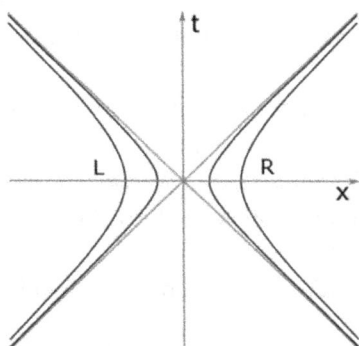

Narozhny et al.,[8] showed that one can attach physical meaning to the Unruh method of quantization only in the double Rindler wedge consisting of L and R, with their regions not being causally connected. The Unruh construction requires that boundary conditions be imposed on the origin (or two-dimensional plane in the case of (3 + 1) dimensional spacetime). These boundary conditions constitute a topological obstacle that disallows any correlation between particles in the L and R wedges. This means that the averaging over quantum field states in one wedge cannot lead to thermalization of states in the other wedge. A free quantum field in Minkowski space cannot be decomposed into two noninteracting fields, one in the L wedge and the other in the R wedge.

At this point, I will quote the very well supported conclusion of Narozhny et al.: "Hence, considerations of the Unruh problem, both in

---

[8] Narozhny et al., *Phys. Rev. D* **65** (2001), 025004.

the standard and algebraic formulations of quantum field theory as of now do not give convincing arguments in favor of a universal thermal response of detectors uniformly accelerated in Minkowski space." Given the intense controversy over the issue, it is worth looking at the Acknowledgements list in this paper.

Hawking Radiation

Although the Unruh and Hawking effects are generally thought to be mathematically and physically distinct, there is a good argument which shows that their relationship through the equivalence principle is closer than one might think.

One often finds Hawking radiation described as the thermal radiation predicted to be spontaneously emitted by black holes, which arises from vacuum fluctuations of particle and antiparticle pairs. One of the particles of the pair is absorbed by the horizon and the other is identified with the Hawking radiation. However, the best way to understand Hawking radiation is to consult Hawking's very clear original 1975 paper.[9]

Numerous authors have discussed the fact that the vacuum and particle states arising in the canonical quantization of the free scalar field depend upon the coordinate system within which the Klein–Gordon equation is solved. The procedure used is to solve the Klein–Gordon equation for the normal modes appropriate for a given curvilinear coordinate system, and relate these to the plane-wave mode solutions in rectilinear Minkowski coordinates via a Bogoliubov transformation.[10] If the coefficient in this transformation corresponding to an admixture of positive and negative frequency modes is non-zero, particles will be present. The mixing of positive and negative frequency modes also implies that the vacuum states will differ. What follows is a brief introduction to these concepts.

---

[9] S. Hawking, *Commun. Math. Phys.* **43** (1975), 199–220.

[10] These transformations were introduced by Bogoliubov in the context of solid-state physics (*Zh. ETF* **34** (1958), 58 [*JETP* **7** (1958), 51]).

A distinction between positive and negative frequency solutions to a general spacetime is possible only if the spacetime possesses a global Killing vector field. This will be assumed to be the case. The generalization of the Klein–Gordon equation to a general spacetime is

$$\Box \psi + m^2 = 0 , \tag{B.1}$$

where

$$\Box = |g|^{-\frac{1}{2}} \partial_\mu \left(|g|^{\frac{1}{2}} g^{\mu\nu} \partial_\nu \right) = g^{\mu\nu} \nabla_\mu \nabla_\nu , \tag{B.2}$$

and ∇ is the covariant derivative. Since the existence of a global time-like Killing vector K is assumed, the normal mode solutions of Eq. (B.2) may be chosen to satisfy

$$\mathcal{L}_K \psi = -iE\psi , \tag{B.3}$$

Where $\mathcal{L}_K$ is the Lie derivative with respect to K. The presence of a Killing vector means that

$$\mathcal{L}_K g_{\mu\nu} = K_{\mu;\nu} + K_{\nu;\mu} = 0 . \tag{B.4}$$

If the coordinate system is chosen such that only the non-zero component of the Killing vector is a unit vector along x^0, then Eq. (B.3) can be written as

$$\partial_{x^0} \psi = -iE\psi , \tag{B.5}$$

and Eq. (B.1) can be solved by separation of variables by the substitution

$$\psi = \psi_j(\vec{x}) e^{-iE_j x^0}. \tag{B.6}$$

If ψ_1 and ψ_2 are complex solutions of Eq. (B.1) and Σ is any complete Cauchy hypersurface for this equation, an inner product can be defined as

$$\langle \psi_1, \psi_2 \rangle = i \int_\Sigma \psi_1^* \overleftrightarrow{f^\mu} \psi_2 \, d\Sigma_\mu \tag{B.7}$$

where,

$$\overleftarrow{f^\mu} = g^{1/2}g^{\mu\nu}\overrightarrow{\partial_{x^\nu}} - \overleftarrow{\partial_{x^\nu}}g^{1/2}g^{\mu\nu} \tag{B.8}$$

and $d\Sigma_\mu$ is the outwardly directed surface element of Σ. The value of $\langle\psi_1,\psi_2\rangle$ is independent of Σ. The arrows above the partial derivatives denote whether the derivative is acting to the left or the right.

The field may then be quantized by defining a field operator[11]

$$\Phi = \sum_i (a_i\,\psi_i + a_i^\dagger\psi_i^*), \tag{B.9}$$

its conjugate momentum $\Pi^\mu = g^{1/2}g^{\mu\nu}\partial_{x^\nu}\Phi$, and imposing the usual commutation relations,

$$[\Phi(x^0,\vec{x}),\Phi(x^0,\vec{x}')] = [\Pi(x^0,\vec{x}),\Pi(x^0,\vec{x}')] = 0$$
$$[\Phi(x^0,\vec{x}),\Pi(x^0,\vec{x}')] = i\delta^3(\vec{x} - \vec{x}')\,. \tag{B.10}$$

Using the definition of Π^μ and Eq. (B.9) gives, when combined with Eq. (B.10), the commutation relations for the annihilation and creation operators a and $a^\dagger$,

$$[a_i,a_j] = [a_i^\dagger,a_j^\dagger] = 0, [a_i,a_j^\dagger] = \delta_{ij} \tag{B.11}$$

Note that a and $a^\dagger$ are operators with no time or space dependence, and ψ_i and ψ_i^* correspond respectively to positive and negative frequency solutions.

Consider now a second set of modes $\bar{\psi}_i$, which in the present context arise from solving the Klein–Gordon equation, (B.1), in some flat-space coordinate system other than rectangular Minkowski coordinates. The new modes $\bar{\psi}_i$ can be expanded in terms of the old modes of Eq. (B.9) as

$$\bar{\psi}_i = \sum_j (\alpha_{ij}\,\psi_j + \beta_{ij}\psi_j^*)\,. \tag{B.12}$$

---

[11] The symbol * is the complex conjugate and † is the adjoint or Hermitian conjugate.

The inner product, Eq. (B.7), can be used to determine the α_{ij} and β_{ij} as

$$\alpha_{ij} = \langle \bar{\psi}_i, \psi_j \rangle, \qquad \beta_{ij} = -\langle \bar{\psi}_i, \psi_j^* \rangle. \tag{B.13}$$

If the field operator Φ is to be expanded in terms of the new modes as

$$\Phi = \sum_i (\bar{a}_i\, \bar{\psi}_i + \bar{a}_i^\dagger \bar{\psi}_i^*), \tag{B.14}$$

the new creation and destruction operators $\bar{a}_i$ and $\bar{a}_i^\dagger$ must be related to the old by

$$\bar{a}_i = \sum_j (\alpha_{ij}^* a_j - \beta_{ij}^* a_j^\dagger), \quad \bar{a}_i^\dagger = \sum_j (\alpha_{ij} a_j^\dagger - \beta_{ij} a_j). \tag{B.15}$$

These equations are known as a Bogoliubov transformation of the operators a and $a^\dagger$.

For Eqs. (B.9) and (B.14) to be consistent, and if $\bar{a}$ and $\bar{a}^\dagger$ are to satisfy the same commutation relations as a and $a^\dagger$, the Bogoliubov coefficients introduced above must satisfy

$$\sum_k (\alpha_{ik} \alpha_{jk}^* - \beta_{ik}\beta_{jk}^*) = \delta_{ij}$$

$$\tag{B.16}$$

$$\sum_k (\alpha_{ik} \beta_{jk} - \beta_{ik}\alpha_{jk}) = 0,$$

or, in equivalent but somewhat redundant matrix notation,

$$\sum_k \begin{pmatrix} \alpha & \beta \\ \beta^* & \alpha^* \end{pmatrix}_{ik} \begin{pmatrix} \alpha^\dagger & -\tilde{\beta} \\ -\beta^\dagger & \tilde{\alpha} \end{pmatrix}_{kj} = \begin{pmatrix} 1 & 0 \\ 0 & 1 \end{pmatrix} \delta_{ij}, \tag{B.17}$$

where the tilde corresponds to matrix transposition.

One can now define the particle number operator, which is the sum of the particle-number operator for each of the states. It is possible to find a common set of eigenstates for these commuting operators, each of which are fully characterized by specifying the particle numbers. In this way, one can form the basis of the particle number representation of the Hilbert space, sometimes called the Fock space.

However, since there are two vacuum states, $|0\rangle$ and $|\bar{0}\rangle$, where $a_i|0\rangle = 0$ and $\bar{a}_i|\bar{0}\rangle = 0$ for all i, two Fock spaces are necessary and they will differ if $\beta_{ij} \neq 0$. This can be seen by direct computation of the matrix element $\bar{N}_i$:

$$\bar{N}_i = \langle 0|\bar{a}_i^\dagger \bar{a}_i|0\rangle = \sum_j |\beta_{ij}|^2, \tag{B.18}$$

which can be written, by summing over i, as the number operator, $\bar{N}$,

$$\bar{N} = \sum_i \langle 0|\bar{a}_i^\dagger \bar{a}_i|0\rangle = Tr\beta\beta^\dagger. \tag{B.19}$$

$\bar{N}_i$ is interpreted as the average number of $\bar{\psi}_i$-mode particles in the vacuum state $|0\rangle$. Note that if $\bar{N}$ diverges, the two vacuum states $|0\rangle$ and $|\bar{0}\rangle$ are not related by a unitary transformation.

The Underlying Physics of Hawking Radiation

Over twenty-five years ago Punsly[12] proposed that a global quantum field theory such that in a Schwarzschild background space when restricted to any point in that space is consistent with the field theory found by a freely falling observer at that point. This can be expanded to small regions of spacetime much less than the radius of curvature.

The equivalence principle demands that the field theory found by a freely falling observer be locally the same as that in flat spacetime. Punsly's approach predicts that an isolated black hole will emit thermal radiation that can be identified with Hawking radiation. He showed that the renormalized stress-energy tensor is a measure of the change in the energy of the zero-point oscillations of the field theory as formulated by

---

[12] B. Punsly, "Black-hole evaporation and the equivalence principle," *Phys. Rev.* D **46** (1992), 1288–1311.

a freely falling observer; an observer at infinity sees the zero-point energy decrease as this observer approaches the black hole horizon. The freely falling observer sees no particles in the local vacuum; i.e., Hawking radiation does not exist for a freely falling observer.

• • •

What is shown above is that both the Unruh effect and Hawking radiation depend on the reality of vacuum fluctuations as defined in the first part of this Appendix. Because there is as yet no truly compelling evidence that these fluctuations actually exist, the Unruh effect and Hawking radiation may well be illusory.

Charge, Spacetime Geometry, and Effective Mass[13]

When one thinks of solutions to the Einstein gravitational field equations, it is often thought that the solutions only have a positive curvature of spacetime associated with them. But this is not always true even for spherically symmetric non-rotating solutions or those having angular momentum. This section shows how negative curvature arises due to electric charge.

Charge, like mass in Newtonian mechanics, is an irreducible element of electromagnetic theory that must be introduced *ab initio*. Its origin is not really a part of the theory. Fields are then defined in terms of forces on either mass — as in the case of Newtonian mechanics, or charges in the case of electromagnetism. General Relativity changed our way of thinking about the gravitational field by replacing the concept of a force field with the curvature of spacetime. Mass, however, remained an irreducible element. It is shown here that the Reissner–Nordström solution to the Einstein field equations tells us that charge, like mass, has a unique spacetime signature.

The Reissner–Nordström solution is the unique, asymptotically flat, and static solution to the spherically symmetric Einstein–Maxwell field equations. Its accepted interpretation is that of a charged mass charac-

---

[13] This section originally appeared in: G. E. Marsh, *Found. Phys.* **38** (2008), 293–300.

terized by two parameters, the mass M and the charge q. While this solution[14] has been known since 1916, there still remains a good deal to be learned from it about the nature of charge and its effect on spacetime.

It will be shown here that if the source of the field is the singularity of the vacuum Reissner–Nordström solution, only the Schwarzshild mass is seen at infinity, with the charge and its electric field making no contribution. In particular, if the charge alone is the source of the field, the effective mass seen at infinity vanishes. This is not the case when the source of the field is a "realistic" source characterized by a mass and proper charge density. It will also be seen that the presence of charge results in a negative curvature of spacetime.

The Reissner–Nordström solution is given by

$$ds^2 = -\left(1 - \frac{2m}{r} + \frac{Q^2}{r^2}\right)dt^2 + \left(1 - \frac{2m}{r} + \frac{Q^2}{r^2}\right)^{-1} dr^2$$

$$+ r^2(d\theta^2 + \sin^2\theta \, d\phi^2), \tag{B.20}$$

where $m = GM/c^2$ and $Q = (G^{1/2}/c^2)q$. The Reissner–Nordström metric reduces to that of Schwarzschild for the case where $Q = 0$. Notice that this metric takes the Minkowski form when $r = Q^2/2m$.

If $Q^2 \leq m^2$, this solution has two apparently singular surfaces located at $r_\pm = m \pm (m^2 - Q^2)^{\frac{1}{2}}$. These are coordinate singularities that may be removed by choosing suitable coordinates and extending the manifold. If $Q^2 = m^2$, these surfaces coalesce into a single surface located at $r = m$, and if $Q^2 > m^2$ the metric is non-singular everywhere except for the origin. These singular surfaces play no role in what follows. An extensive discussion of the vacuum Reissner–Nordström and Schwarzschild solutions, along with their Penrose diagrams was given by Hawking and Ellis.[15]

---

[14] H. Reissner, "Über die eigengravitation des elektrischen feldes nach der Einstein'schen Theorie," *Ann. Physik* **50** (1916), 106–120; G. Nordström, "On the energy of the gravitational field in Einstein's theory," *Verhandl. Koninkl. Ned. Akad. Wetenschap., Afdel. Natuurk., Amsterdam* **26** (1918), 1201–1208.

[15] S. W. Hawking and G. F. R. Ellis, *The Large Scale Structure of Space-Time* (Cambridge University Press, Cambridge, 1973), pp. 156–161.

Most applications of the Reissner–Nordström solution would be outside a body responsible for the charge and mass. Here it is the *vacuum* solution to the field equations, considered to be valid for all values of r, that is of interest.

Like the vacuum Schwarzschild solution, the Reissner–Nordström vacuum solution has an irremovable singularity (in the sense that it is not coordinate dependent) at the origin representing the source of the field. In what follows, only the Reissner–Nordström solution having this singularity as a source of the field will be considered.

The interesting thing about the singularity is that it is time-like so that clocks near the singularity run *faster* than those at infinity. It is also known that the singularity of the Reissner–Nordström solution is repulsive in that time-like geodesics will not reach the singularity.

Curvature in the Reissner–Nordström Solution

If one computes the Gaussian curvature associated with the Schwarzschild solution, it is readily seen that the curvature vanishes. Higher order scalars, such as the Kretschmann scalar given by $K = R_{\alpha\beta\gamma\delta}R^{\alpha\beta\gamma\delta}$, do not vanish, but their interpretation is problematic.[16] Of course, the curvature of spacetime around a Schwarzschild black hole does not vanish since the curvature tensor does not vanish. More important for the present discussion is that a simple way to determine the sign of the curvature is well known.

Consider first the Schwarzschild solution. Draw a circle on the equatorial plane where $\theta = \pi/2$ is centered on the origin. The circumference of this circle is $2\pi r$. The proper radius from the origin to the circle is given by

$$\int_0^r \sqrt{g_{11}}dr = \int_0^r \left(1 - \frac{2m}{r}\right)^{-1/2} dr \geq r.$$ (B.21)

---

[16] The divergence of the Kretschmann scalar as $r \to 0$ indicates a real — as opposed to a coordinate dependent — singularity. It has been proposed that the Kretschmann scalar be called "the spacetime curvature" of a black hole; see: R. C. Henry, "Kretschmann scalar for a Kerr–Neuman Black Hole", *Astrophys. J.* **535** (2000), 350–353.

Consequently, the ratio of the circumference of the circle to the proper radius is less than or equal to 2π. This tells us that the space is positively curved. Now consider a *negative* mass. The inequality sign in Eq. (B.21) reverses so that the ratio of the circumference of a circle to its proper radius is greater than 2π, telling us that the space surrounding a negative mass has a negative curvature.

The case of the Reissner–Nordström solution is more interesting. Setting

$$g_{00} = \left(1 - \frac{2m}{r} + \frac{Q^2}{r^2}\right) \text{ and } g_{11} = \left(1 - \frac{2m}{r} + \frac{Q^2}{r^2}\right)^{-1},$$

and using the above method of determining the spatial curvature gives the results shown in Table 1. For $r < Q^2/2m$, one has a *negatively* curved spacetime, which is embedded in a positively curved spacetime with a $(2 + 1)$ dimensional boundary having the Minkowski form between them. In the region between the time-like singularity at the origin and the $(2 + 1)$ dimensional hypersurface, the spacetime is negatively curved independent of the sign of the charge. This implies that charge manifests itself as a negative curvature — just as mass causes a positive curvature.

Table B.1. The metric coefficients g_{00} and g_{11} for different ranges of r, and the sign of the spatial curvature in these regions.

	$r > Q^2/2m$	$r = Q^2/2m$	$r < Q^2/2m$
g_{00}	≥ -1	-1	< -1
g_{11}	> 1	1	< 1
Spatial Curvature	Positive	Flat	Negative

That charge effectively acts as a negative mass can also be seen from the equations governing the motion of a test particle near a Reissner–Nordström singularity. For an uncharged particle falling inward towards the singularity the radial acceleration is,[17]

$$\frac{d^2r}{d\tau^2} = -\frac{1}{r^2}\left(m - \frac{Q^2}{r}\right).$$ (B.22)

The gravitational field that affects the test particle varies with distance from the singularity and becomes repulsive when the effective mass $m_{\text{eff}} = \left(m - \frac{Q^2}{r}\right)$ becomes negative at $r < Q^2/m$. Neutral matter falling into the singularity would therefore ultimately accumulate on the $(2+1)$-dimensional spherical hypersurface where $m_{\text{eff}} = 0$.

Thus, by means of very straight-forward considerations, the Reissner–Nordström solution leads to the conclusion that charge — of either sign — causes a negative curvature of spacetime.

The Electric Field

This section is devoted to a general relativistic calculation of the effective mass of the vacuum Reissner–Nordström solution: first, of that contained within the interior of a spherical surface of radius R, centered on the singularity — and designated $M_{\text{Eff}}^{\text{In}}$; and second, the effective mass of the electric field alone outside that surface — designated $M_{\text{Eff}}^{\text{Out}}$. The key references for what follows are Synge,[18] and Gautreau and Hoffman.[19]

---

[17] V. de la Cruz and W. Israel, "Gravitational bounce," *Nuovo Cimento* **51** (1967), 744; J. M. Cohen and D. G. Gautreau, "Naked singularities, event horizon, and charged particles," *Phys. Rev. D* **19** (1979), 2273–2279; W. A. Hiscock, "On the topology of charged spherical collapse," *J. Math. Phys.* **22** (1981), 215; F. de Felice and C. J. S. Clarke, *Relativity on Curved Manifolds* (Cambridge University Press, Cambridge, 1992), pp. 369–372.

[18] J. L. Synge, *Relativity: The General Theory* (North-Holland Publishing Company, Amsterdam, 1966), Ch. VII, §5 and Ch. X, §4.

[19] R. Gautreau and R. B. Hoffman, "The structure of the sources of Weyl-type electrovac fields in general relativity," *Nuovo Cimento* **16** (1973), 162–171.

Synge gives the following Stokes relation[20] for a three-dimensional volume, v_3, bounded by a closed two-surface v_2:

$$\oint_{v_2} V_{,i}\, n^i dv_2 = \frac{1}{2}\int_{v_3} (G_4^4 - G_i^i)V dv_3 . \qquad (B.23)$$

Here, dv_2 and dv_3 are the invariant elements of area and volume, G is the Einstein tensor, and V is defined by the line element

$$ds^2 = g_{ij}dx^i dx^j - V^2 dt^2 , \qquad (B.24)$$

which, at infinity, is assumed to take the form of the Minkowski metric. n^i is the outward unit normal to the surface v_2. Einstein's equations, $G_{\mu\nu} = -\kappa T_{\mu\nu}$, with $\kappa = 8\pi$, allow Eq. (B.23) to be written as

$$\oint_{v_2} V_{,i}\, n^i dv_2 = 4\pi \int_{v_3} (T_i^i - T_4^4)V dv_3 . \qquad (B.25)$$

The integral on the right-hand side of this equation corresponds to the *total effective mass* enclosed by the surface v_2. This is known as Whittaker's theorem.[21] Thus,

$$M_{\text{Eff}}^{\text{In}} = \frac{1}{4\pi}\oint_{v_2} V_{,i}\, n^i dv_2 . \qquad (B.26)$$

Note that the effective mass, as defined by Eqs. (B.25) and (B.26), depends only on the energy-momentum tensor and the g_{00} component of the metric. Choose a spherical surface of radius R with the Reissner–Nordström singularity at the origin. From Eq. (B.20), V is given on the surface as

$$V = \left(1 - \frac{2m}{R} + \frac{Q^2}{R^2}\right)^{1/2} , \qquad (B.27)$$

---

[20] Greek indices take the values 1, 2, 3, 4 and Latin indices 1, 2, 3. To avoid unnecessary confusion, the notation used here is generally consistent with that found in the relevant literature.

[21] E. T. Whittaker, "On Gauss' theorem and the concept of mass in general relativity," *Proc. Roy. Soc. London* **A149** (1935), 384.

$dv_2 = R^2 \sin\theta \, d\theta d\varphi$ and $n^i = (V, 0, 0)$. Substituting into Eq. (B.26) gives the result quoted above [just after Eq. (B.22)] for m_{eff} at a distance R from the singularity

$$M_{\text{Eff}}^{\text{In}} = m - \frac{Q^2}{R}. \tag{B.28}$$

For asymptotically flat spacetimes, global quantities such as the total energy can be defined as surface integrals in the asymptotic region. This is the basis for the definition of the ADM energy (or mass).[22] What will be shown here is that for $R \neq \infty$, the sum of the effective mass within the surface v_2 and that exterior to v_2 is the Schwarzschild mass. This is true for the vacuum solution being considered here, not necessarily for realistic sources such as those considered by Cohen and Gautreau.

Whittaker's theorem allows the effective mass enclosed by the surface v_2, which is composed of the mass located at the origin and that corresponding to the electric field within v_2, to be written as in Eq. (B.28).

One can also compute the effective mass exterior to the surface v_2. There, the only energy density to be found is that associated with the electric field. By summing the effective mass found in the volumes both interior and exterior to v_2, one obtains the effective mass enclosed by the surface at infinity; that is, the ADM mass. Given that global quantities defined by surface integrals in the asymptotic region cannot generally be written as volume integrals over the interior region, this is a somewhat surprising result.

How to use the relation of Eq. (B.25) to compute the electric field energy in the volume *exterior* to the spherical surface of radius R centered on the singularity can be understood by referring to Fig. B.1. The volume of interest is v_3' *exterior* to the surface v_2. It has two boundary components, the "surface at infinity" and v_2 itself.

---

[22] R. Arnowitt, S. Deser, and C. W. Misner, "The dynamics of general relativity," contained in L. Witten (ed.), *Gravitation: An Introduction to Current Research* (John Wiley & Sons, Inc., New York, 1962), pp. 227–265.

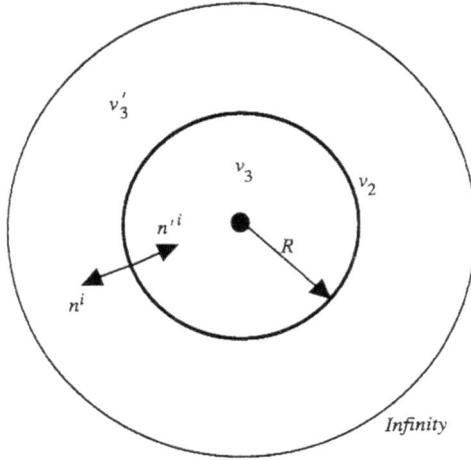

Fig. B.1. The Reissner–Nordström singularity is located at the center of the spherical surface v_2 of radius R enclosing the volume v_3. The outward pointing unit normal to v_2 is n^i. The surface enclosing the volume v_3' is composed of the point at infinity and v_2. The outwardly pointing unit normal to v_2, when acting as a boundary component of v_3', is n'^i.

Since the surface integral at infinity vanishes, Eq. (B.25) for the volume v_3' may be written as

$$4\pi \int_{v_3'} (T_i^i - T_4^4) V' dv_3' = \oint_{v_2} V_{,i}' \, n'^i dv_2 . \tag{B.29}$$

The V in Eq. (B.25) has been changed to V' in Eq. (B.29). The reason for this is that the energy-momentum tensor in the volume v_3' must be restricted to the contribution from only the electric field since no masses exist in v_3'. The way to do this is to recognize that the Reissner–Nordström solution remains a solution to the Einstein field equations even when the mass m is set equal to zero. The resulting metric is that for a massless point charge, which — as discussed above — has a negative curvature and is repulsive. The energy-momentum tensor for the electric field nonetheless has a positive energy density. The V' that should be used in Eq. (10) is therefore that from the metric for a massless point charge; i.e.,

$$V' = \left(1 + \frac{Q^2}{r^2}\right)^{1/2} \tag{B.30}$$

Note that r takes the fixed value R when computing the surface integral.

Because $n^i = -n'^i$, the effective mass contained in the volume v'_3 exterior to v_2 is

$$M_{\text{Eff}}^{\text{Out}} = \frac{1}{4\pi} \oint_{v_2} V'_{,i} \, n'^i dv_2$$

$$= -\frac{1}{4\pi} \oint_{v_2} V'_{,i} \, n^i dv_2 = -\int_R^{\infty} (T_i^i - T_4^4) V' dv'_3. \tag{B.31}$$

$M_{\text{Eff}}^{\text{Out}}$ can be evaluated by simply using the second integral in Eq. (B.31), which was already evaluated for V above. Taking account of the orientation of the surface and the substitution of V', the result is

$$M_{\text{Eff}}^{\text{Out}} = \frac{Q^2}{R}. \tag{B.32}$$

Combined with Eq. (B.29), this results in

$$M_{\text{Eff}}^{\text{In}} + M_{\text{Eff}}^{\text{Out}} = m. \tag{B.33}$$

What this tells us is that the "negative mass" associated with the charge Q [see Eq. (B.28)] is exactly compensated by the effective mass contained in the electric field present in the volume exterior to the surface $r = R$. If the radius $r = R \to \infty$, the effective mass contained within the surface at infinity is m, the Schwarzschild or equivalently, the ADM mass.

One can also obtain the result given in Eq. (B.33) by directly evaluating the last integral on the right-hand side of Eq. (B.31). This will be done here for the sake of completeness as well as a confirmation of Eq. (B.32) above. To begin with, an identity relating the energy-momentum tensor F^{α}_{β} of the electric field to the scalar potential is needed.

If the energy-momentum tensor

$$T_{\mu\upsilon} = F_\mu^\alpha F_{\alpha\upsilon} - \frac{1}{4} g_{\mu\upsilon} F_{\alpha\beta} F^{\alpha\beta} , \tag{B.34}$$

where

$$F_{\mu\upsilon} = \frac{\partial A_\upsilon}{\partial x^\mu} - \frac{\partial A_\mu}{\partial x^\upsilon}$$

is restricted to the case where only electric fields are present, so that

$$A_\mu = (0, 0, 0, \frac{1}{\sqrt{4\pi}} \phi , \tag{B.35}$$

then it is readily shown that

$$F_{i4} = \frac{1}{\sqrt{4\pi}} \phi_{,i} \quad \text{and} \quad F^{i4} = -\frac{1}{\sqrt{4\pi}} (V')^{-2} g^{ij} \phi_{,j} . \tag{B.36}$$

Equations (B.36) allow the energy-momentum tensor to be written as

$$4\pi T_{ij} = \frac{2}{V^2} \left(\frac{1}{2} g_{ij} \Delta_1 \phi - \phi_{,i} \phi_{,j} \right) , \tag{B.37}$$

where Δ_1 is a differential parameter of the first order defined[23] by

$$\Delta_1 \phi = g^{ij} \phi_{,i} \phi_{,j} . \tag{B.38}$$

The needed identity may now be obtained from Eq. (B.37) as

$$4\pi(T_i^i - T_4^4) = \frac{\Delta_1 \phi}{V'^2} , \tag{B.39}$$

which, for spherical coordinates, may be written as

$$4\pi(T_i^i - T_4^4) = \frac{g^{11}(\phi_{,r})^2}{V'^2} . \tag{B.40}$$

---

[23] L. P. Eisenhart, *Riemannian Geometry* (Princeton University Press, Princeton, 1997), p. 41.

The total effective mass inside the three-volume dv_3' is then

$$M_{\text{Eff}}^{\text{Out}} = -\int_R^\infty (T_i^i - T_4^4)V'dv_3' = -\frac{1}{4\pi}\int_R^\infty \frac{g^{11}(\phi_{,r})^2}{V'}dv_3'. \quad (B.41)$$

Substitution of V' from Eq. (B.30), along with $g^{11} = \left(1 + \frac{Q^2}{r^2}\right)$, $\phi = Q/r$, and $dv_3' = \frac{r^2}{V'}\sin\theta\, d\theta d\varphi$, yields

$$M_{\text{Eff}}^{\text{Out}} = -\int_R^\infty \frac{Q^2}{r^2}dr = \frac{Q^2}{R}. \quad (B.42)$$

As expected, this is the same result as that given in Eq. (B.32).

Summary

The above results may then be summarized as in Eq. (B.33)

$$M_{\text{Eff}}^{\text{In}} + M_{\text{Eff}}^{\text{Out}} = m$$

independent of the radius R. What this says is that the amount of "negative mass" due to the term $-Q^2/R$ in Eq. (B.28) is exactly compensated by the amount of "positive mass" contained in the region $r > R$. For R infinite, $M_{\text{Eff}}^{\text{In}}$ is the Schwarzschild mass; and if $R < \infty$, $M_{\text{Eff}}^{\text{In}}$ is less than the Schwarzschild mass.

In their 1979 paper, Cohen and Gautreau noted that: "As R decreases, M_T [here equal to $M_{\text{Eff}}^{\text{In}}$] also decreases because the electric field energy inside a sphere of radius R decreases." And, one might add, as R decreases, the field energy exterior to R increases. This is equivalent to

$$M_{\text{Eff}}^{\text{In}} = \left(m - \frac{Q^2}{R}\right) \quad \Rightarrow \quad \frac{dM_{\text{Eff}}^{\text{In}}}{dR} = \frac{Q^2}{R^2}$$

$$M_{\text{Eff}}^{\text{Out}} = \frac{Q^2}{R} \quad \Rightarrow \quad \frac{dM_{\text{Eff}}^{\text{Out}}}{dR} = -\frac{Q^2}{R^2}, \quad (B.43)$$

so that

$$\frac{dM_{\text{Eff}}^{\text{In}}}{dR} + \frac{M_{\text{Eff}}^{\text{Out}}}{dR} = 0 \,. \tag{B.44}$$

While charge of either sign causes a negative curvature of spacetime, the Einstein–Maxwell system of equations does not allow different geometric representations for positive and negative charges. This is a direct result of the fact that the sources of the Einstein–Maxwell system are embodied in the energy-momentum tensor, which depends only on the (non-gravitational) energy density — which is why charge enters as Q^2 above. Thus, a full geometrization of charge does not appear to be possible within the framework of the Einstein–Maxwell equations.

As mentioned earlier, no "realistic" sources for the Reissner–Nordström metric are considered in this paper. Realistic sources raise many interesting questions, among them are: Can a lone, charged black hole actually exist? If so, how can global charge neutrality be maintained?

•   •   •

The above applied to the vacuum Reissner–Nordström metric, but the work can be extended to charged rotating vacuum solutions of the Einstein field equations, and in particular, to the Kerr–Newman solution.[24]

The Kerr–Newman solution in generalized Eddington coordinates,[25] which are convenient for this approach, is given by

$$ds^2 = dr^2 - 2a \sin^2\theta \, drd\phi + (r^2 + a^2)\sin^2\theta \, d\phi^2$$
$$+ (r^2 + a^2\cos^2\theta)d\theta^2 - dt^2 + \frac{2mr - Q^2}{r^2 + a^2\cos^2\theta}$$
$$- [dr - a \sin^2\theta \, d\phi + dt]^2 \,, \tag{B.45}$$

where the symbols have their conventional meanings.

---

[24] G. E. Marsh, "Charge geometry and effective mass in the Kerr-Newman solution to the Einstein field equations," *Found. Phys.* **38** (2008) 959–968.

[25] An excellent discussion of these coordinates and their interpretation can be found in R. H. Boyer and R. W. Lindquist, "Maximal analytic extension of the Kerr metric," *J. Math. Phys.* **8** (1967), 265–281.

It was pointed out above that for the Reissner–Nordström solution the metric takes the Minkowski form when $r = Q^2/2m$. Interestingly enough, the same thing occurs in the Kerr–Newman metric except that now r has a different meaning with surfaces of constant r corresponding to confocal ellipsoids satisfying

$$\frac{x^2 + y^2}{r^2 + a^2} + \frac{z^2}{r^2} = 1. \tag{B.46}$$

It will be seen, however, that unlike the Reissner–Nordström solution, where it was possible to show that for $r < Q^2/2m$ the curvature was negative, the case of the Kerr–Newman solution is more complex. An indication of this is given in Fig. B.2.

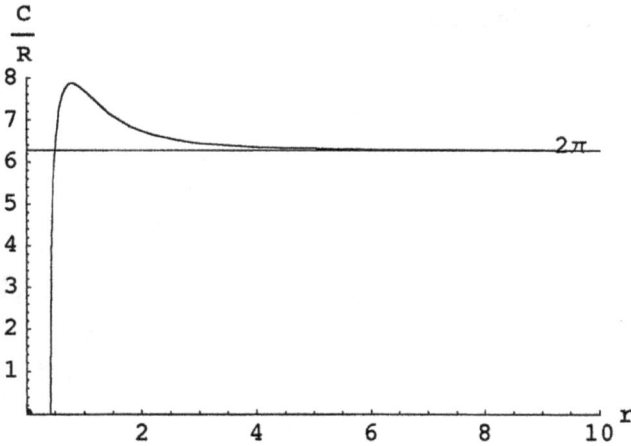

Fig. B.2. The ratio of the circumference to the radius R in the equatorial plane of the Kerr–Newman solution in Eddington coordinates. The ring singularity of the Kerr–Newman metric is at $r = 0$. The portion of the curve above the line $C/R = 2\pi$ corresponds to a negative curvature and that below to positive curvature. $C/R = 0$ at $r \sim 0.404698$ where $g_{\phi\phi} = 0$, and crosses the line $C/R = 2\pi$ at $r = Q^2/2m$, which for the value of the parameters used here, $a = m = Q = 1$, is 0.5.

Here the negative contribution of rotation to the effective mass for the Kerr–Newman metric has been excluded. The "negative mass" due to charge has properties very similar to that of the Reissner–Nordström metric. Both take the Minkowski form at $r = Q^2/2m$, even though the meaning of r is different for the two metrics; the effective mass interior to this surface is $-m$ in both cases; and both have an effective mass of m at infinity. In addition, the effective mass for both metrics satisfies, for any surface defined by $r = $ constant (again for either definition of r), the relation

$$M_{\text{Eff}}^{\text{Int}} + M_{\text{Eff}}^{\text{Ext}} = m \tag{B.47}$$

Thus, the positive effective mass of the electric field exterior to the surface exactly compensates for the "negative mass" associated with the charge located within the surface.

Appendix C

The Vacuum and the
Cosmological Constant Problem

Introduction

The cosmological constant problem exists because of a number of key assumptions. These will be identified in this introduction. The second section discusses the issue of vacuum instability and negative energy states. Following this is a section on the relation between gauge invariance, Schwinger terms, and the definition of the vacuum. Given recent observational data from the Supernova Cosmology Project — showing an acceleration of the expansion of the universe at great distances, it is attractive to look at definitions of the vacuum that could lead to reasonable, but non-zero values of the cosmological constant.

While it has never caught a great deal of interest, the redefinition of the vacuum proposed by Solomon[1] is one possible approach to achieving this goal. It is included here primarily for heuristic purposes.

The final section looks at the origin of the difficulties in QFT and also serves as a summary.

Perhaps the best introduction to the cosmological constant problem is the review by Weinberg[2] published over twenty years ago. Although there have been a variety of approaches to the issue since then, the article can still serve to frame the problem, at least in the sense that it is to be addressed here. To avoid confusion, Weinberg's notation will be used in this introduction.

---

[1] D. Solomon, "Gauge invariance and the vacuum state," *Can. J. Phys.* **76** (1998), 111–127.

[2] S. Weinberg, "The cosmological constant problem," *Rev. Mod. Phys.* **61** (1989), 1–23.

The Einstein field equations, including the cosmological constant λ, are

$$R_{\mu\nu} - \frac{1}{2}g_{\mu\nu}R - \lambda g_{\mu\nu} = -8\pi G T_{\mu\nu} \,. \tag{C.1}$$

The right-hand side of the equation contains the energy-momentum tensor $T_{\mu\nu}$, and it is here that the first assumption that leads to the cosmological constant problem is made. It is assumed that the vacuum has a non-zero energy density. If such a vacuum energy density exists, Lorentz invariance requires that it have the form[3]

$$\langle T_{\mu\nu}\rangle_{\text{Vac}} = -\langle\rho\rangle g_{\mu\nu} \,. \tag{C.2}$$

This allows one to define an effective cosmological constant and a total effective vacuum energy density

$$\lambda_{\text{eff}} = \lambda + 8\pi G\langle\rho\rangle$$
$$\rho_V = \langle\rho\rangle + \frac{\lambda}{8\pi G} = \frac{\lambda_{\text{eff}}}{8\pi G} \,. \tag{C.3}$$

Observationally, it is known that the total effective vacuum energy density must be comparable to about $\rho_V \sim 10^{-29}\frac{g}{cm^3}$.

The second assumption that leads to the cosmological constant problem is that the vacuum energy density is due to the zero-point energy of a quantized field or fields. The concept of the zero-point energy originated with the quantization of the simple harmonic oscillations of a particle with non-vanishing mass. This energy is present for each energy level including the ground state. In QFT, at a given instant of time, the field is defined at each point in space — essentially an independent harmonic oscillation at each point with amplitude and phase depending on the initial conditions. By specifying an equation of motion depending on the amplitude of the field and its partial derivative with respect to time, one couples these otherwise independent oscillations. In the Klein–Gordon equation, for example, this coupling is achieved by the presence of the Laplacian.

---

[3] The signature of the metric is $+2$.

The zero-point energy is also present in the relativistic QFT of the neutral Klein–Gordon field and the electromagnetic field. Of course, the zero-point energy of the Klein–Gordon field is infinite, but when used to formulate the cosmological constant problem, it is generally summed up to a cutoff that determines the magnitude of the vacuum energy density. The way this is done is to count the number of normal modes between v and $v + dv$, multiply by the zero-point energy in each mode, and integrate up to the chosen cutoff. There is some question as to whether it makes sense to count the normal modes for a massive particle when addressing the energy density of the vacuum where no particles are present. The usual justification for this is somewhat confused in the literature, but is generally based on identifying the zero-point energy with vacuum fluctuations.

The number of normal modes between v and $v + dv$ is given by

$$dZ = \frac{4\pi V}{c^3} v^2 dv .\tag{C.4}$$

Transforming to wave number, setting the volume equal to unity, and using units where $\hbar = c = 1$, and integrating up to a wave number cutoff $\Lambda \gg 1$ results in a vacuum energy density of

$$\langle \rho \rangle = \int_0^\Lambda \frac{1}{2}\sqrt{k^2 + m^2}\, \frac{4\pi k^2}{(2\pi)^3}\, dk \simeq \frac{\Lambda^4}{16\pi^2} .\tag{C.5}$$

As a relativistic wave equation, the Klein–Gordon equation has an energy spectrum that includes both positive and negative energies, corresponding to either sign for the radical in Eq. (C.5). Note that if the integral were performed for both signs and the results added, the vacuum energy density would vanish. In QFT, zero-point energies are inherent in the canonical field quantization method because the ordering of operators in the Hamiltonian is not fixed.

In Eq. (C.5), the vacuum energy density up to a cutoff is calculated for the cosmological constant problem by using only the positive sign for the radical. The reason for this goes back to Dirac's redefinition of the vacuum where electrons are assumed to fill the negative-energy states, and they and their infinite vacuum charge density are presumed to be unobservable — as are the negative-energy state zero-point energies. In this way, the instability of the vacuum caused by the availability of negative energy states is eliminated. The normal ordering procedure of canonical QFT, in addition to eliminating the zero-point energies, also has the virtue of eliminating the infinite vacuum charge density of the Dirac vacuum. QFT does not, however, eliminate the problem of negative energies as will be seen below.

The value of wave number cutoff to be chosen in Eq. (C.5) depends on one's view of Einstein's theory of gravitation. Fields are generally defined in the context of a background metric for spacetime — the Minkowski metric. General relativity has to do with the geometry of spacetime itself, and while there is no *a priori* reason (or experimental evidence) that the gravitational field need be quantized, it is generally believed that Einstein's theory of gravitation will no longer hold at very small distances, and in particular, for distances comparable or smaller than the Planck length $(G\hbar/c^3)^{1/2}$. At this length scale, quantum fluctuations are thought to be the dominant influence on the local spacetime geometry. For this reason, the Planck length is generally chosen to define the cutoff. Such a cutoff is really an expression of our ignorance of the vacuum and the structure (if there is one) of spacetime itself at these distances. Unfortunately, no experimental data exists.

Evaluating the right-hand side of Eq. (C.5) with Weinberg's choice of $\Lambda = (8\pi G)^{-1/2}$ (the units are again such that $\hbar = c = 1$) results in a value for the vacuum energy density of

$$\langle \rho \rangle = 2 \times 10^{89} \frac{g}{cm^3}. \tag{C.6}$$

Comparing this with the observational value of $\rho_V \sim 10^{-29}$ g/cm^3 tells us that the two terms in the effective vacuum energy density of Eq. (C.3) must cancel to some 118 decimal places. This fine tuning *is* the cosmological constant problem as it is currently understood.

As discussed above, the Casimir effect, which in QFT is thought to be due to the presence of vacuum fluctuations, is used to argue that the zero-point energies are real. That is, zero-point energies and quantum fluctuations are identified. A key reference for the Casimir effect is the work of Plunien, Müller, and Greiner.[4] Vacuum fluctuations — which represent the terms of the perturbation series for the vacuum expectation value $\langle 0|S|0 \rangle$ of the S-matrix — are generally ignored since they can only produce an overall phase factor. However, in the presence of external sources or boundaries, these fluctuations can no longer be so casually dismissed. Although the vacuum state must be invariant under rotations and translations — and therefore must have zero momentum, angular momentum, and energy — the presence of external sources or boundaries (as in the Casimir effect) breaks these fundamental symmetries, allowing vacuum fluctuations to have effects that are observable.

Moreover, if the energy of the vacuum is defined as the difference of zero-point energies in the presence of boundaries $\partial \Gamma$ of a region Γ and without boundaries, the vacuum energy can be *negative*; that is,

$$E_{\text{vac}}[\partial \Gamma] = E_0[\partial \Gamma] - E_0[0] \,, \tag{C.7}$$

where $E_0[\partial \Gamma]$ is the zero-point energy in the presence of boundaries and $E_0[0]$ is the zero-point energy in their absence.

Because of the absence of boundaries or external sources, performing the kind of calculation given in Eq. (C.5) has raised questions in the literature with regard to the legitimacy of the approach of simply summing free field modes. See, for example, Lamoreaux[5] where it is argued that we do not learn much about the properties of the vacuum of free space through the study of Casimir and related zero-point energy effects.

In the absence of interactions as well as boundaries, one could make the argument that there appears to be no reason not to allow the vacuum to have a negative energy spectrum so that both signs of the energy in evaluating Eq. (C.5) should be used, thereby eliminating the cosmological constant problem. In the presence of interactions the issue becomes more complicated, but — as will be seen below — the availability of a negative energy spectrum is one way to restore gauge invariance.

---

[4] G. Plunien, B. Müller, and W. Greiner, "The Casimir effect," *Phys. Rep.* **134** (1986), 87–193.

[5] S. K. Lamoreaux, "Casimir forces: Still surprising after 60 years," *Physics Today* (February 2007) pp. 40–45.

Vacuum Instability and Negative Energy States

At this point, it might be useful to examine the origin of the concern about vacuum instability in the presence of negative energy states. Consider a free electron in a positive energy state E subjected to a periodic perturbation of frequency ω. It is generally agreed that there is a non-vanishing probability for the electron to make a transition to a state of energy $(E + \hbar\omega)$ or $(E - \hbar\omega)$. If $\hbar\omega > (E + mc^2)$, it is argued that the transition to $(E - \hbar\omega)$ would be to a state of negative energy. Having set the stage, the argument continues by considering a bound state electron in a hydrogen atom. Since the electron is coupled to the electromagnetic field, such a bound state would rapidly make a radiative transition to a negative energy state, with the result that the hydrogen atom would have no stable existence. Worse yet, since the spectrum has no lower bound, there would be no limit to the radiated energy.

There is an objection to the above argument that shows that such transitions could well be forbidden. Under the usual sign convention, quantum states of positive energy evolve in time as $e^{-i\omega t}$. A state of negative energy, $-E$, then evolves as $e^{+i\omega t}$, which corresponds to the transformation $t \rightarrow -t$. But there are two possibilities for a time reversal operator: it can be unitary or anti-unitary — where $t \rightarrow -t$ and one also takes the complex conjugate of states and complex numbers. Under a unitary transformation, used for all other discrete and continuous symmetries, the time-reversed state corresponds to a state of negative energy $-E$. It was Eugene Wigner who introduced the anti-unitary time reversal operator so as to eliminate negative energies.

There are now two arguments that can be given against the transition from a bound state (as in the above argument) to a negative energy state under the periodic perturbation. First, if the Hamiltonian $\mathcal{H}$ governing the transition is to be CPT invariant — as it must if it is to be an acceptable quantum electrodynamics Hamiltonian — it must satisfy $CPT\ \mathcal{H}(x)\ [CPT]^{-1} = \mathcal{H}(-x)$. This will only be the case if the operator CPT is anti-unitary, a consequence of the complex conjugation implicit in the time reversal operator T. But since the transformation from $e^{-i\omega t} \rightarrow e^{+i\omega t}$ corresponds to simply $t \rightarrow -t$, we have instead

$CPT \mathcal{H}(x) [CPT]^{-1} = -\mathcal{H}(-x)$, giving the negative energy state. Thus, the Hamiltonian of the transition cannot satisfy CPT invariance.

The second argument has to do with the fact that if an electron in an external field obeys the (quantized) Dirac equation, one cannot rule out the negative energy solutions needed to make up a complete set of wavefunctions. But a wavefunction representing a negative energy state can only be non-zero if it has charge $+e$.[6] This means the argument given above for the transition of an electron to a negative energy state will violate charge conservation.

Thus, a radiative transition of an electron to a negative energy state either violates CPT invariance or the conservation of charge. So, it would appear that such transitions are effectively forbidden. This means there is no reason not to allow negative energies to be summed over in Eq. (C.5), yielding a vanishing vacuum energy density. The attractiveness of Solomon's approach to redefining the vacuum, to be described below, is that it allows for incomplete cancellation of the zero-point energies leading to a small, but not vanishing vacuum energy density.

QFT and Gauge Invariance

The issue of the gauge invariance of QFT has been dealt with in a variety of ways over the years (an extensive discussion is contained in Solomon's paper). In essence, the standard vacuum of QFT is only gauge invariant if non-gauge invariant terms are removed. There are two general approaches to the problem: the first is to simply ignore such terms as being physically untenable and remove them so as to maintain gauge invariance; and the second is to use various regularization techniques to cancel the terms. Pauli–Villars regularization[7] is discussed below.

---

[6] S. Weinberg, *The Quantum Theory of Fields* (Cambridge University Press, New York, 1995), Vol. 1, Ch. 14.

[7] W. Pauli and F. Villars, "On the invariant regularization in relativistic quantum theory," *Rev. Mod. Phys.* **21** (1949), 434–444. Because Pauli–Villars regularization does not preserve gauge invariance in non-Abelian gauge theories, dimensional regularization has become the means of choice.

Returning to Eq. (C.5), another problem with this equation is that the introduction of a momentum-space cutoff destroys translational invariance and may make it difficult to maintain gauge and Lorentz invariance. If, instead of introducing a cutoff, an attempt is made to deal with this infinite integral by Pauli–Villars regularization, the result is the introduction of negative masses. This can be seen as follows. It can be guaranteed that the vacuum expectation value of the energy momentum tensor is Lorentz invariant and hence proportional to $\eta^{\mu\nu}$ if we use the relativistic formulation

$$\langle 0|T^{\mu\nu}|0\rangle = \frac{1}{2} \int_0^\infty \frac{1}{(2\pi)^3} \frac{p^\mu p^\nu}{p^0} dp^3 . \tag{C.8}$$

Here $p^0 = E(\boldsymbol{p}) = (\boldsymbol{p}^2 + m^2)^{1/2}$. The domain of integration can be transformed to spherical coordinates in a space of any dimension by the use of the formula

$$\prod_{i=1}^{d} dp^i = p^{d-1} dk \prod_{i=1}^{d-1} \sin^{i-1}\theta_i \, d\,\theta_i , \tag{C.9}$$

resulting in

$$\langle 0|T^{00}|0\rangle = \frac{1}{2} \int_0^\infty \frac{4\pi}{2} \frac{1}{(2\pi)^3} p^2 \, dp \, \sqrt{p^2 + m^2} \tag{C.10}$$

The Pauli–Villars regulator masses m_i, where $m_1 = m$, and associated coefficients c_i, where $c_1 = 1$, are then introduced as follows:

$$\langle 0|T^{00}|0\rangle = \frac{1}{2} \int_0^\infty \frac{4\pi}{2} \frac{1}{(2\pi)^3} p^2 \, dp \sum_i c_i \, p \sqrt{1 + \frac{m^2}{p^2}} . \tag{C.11}$$

The number of regulator masses needed depends in general on the integral. Since we are interested in the convergence of the integral at the upper limit where $m_i^2/p^2 < 1$, the radical can be expanded in a series to yield

$$\langle 0|T^{00}|0\rangle = \frac{1}{2}\int_0^\infty \frac{4\pi}{2}\frac{1}{(2\pi)^3}p^2\,dp \sum_i c_i\left(p + \frac{1}{2}\frac{m^2}{p} - \frac{m^4}{8p^3} + \dots\right).$$

(C.12)

This integral will converge at the upper limit provided the following relations are satisfied as $m_i \to \infty$:

$$1 + \sum_{j=2}^n c_j = 0,$$

(C.13)

$$m^2 + \sum_{j=2}^n c_j m_j^2 = 0, \quad m^4 + \sum_{j=2}^n c_j m_j^4 = 0, \quad m_j \to \infty.$$

Moving p^2 in the numerator of Eq. (C.12) into the series expansion shows that Eqs. (C.13) will be satisfied if $n = 3$.

Because some of the coefficients must be negative, this procedure introduces negative energies in the form of negative masses that are allowed to become infinite at the end of the calculation. The advantage, at least for QFT, is that gauge and Lorentz invariance are preserved.

If one works in Euclidean space by first performing a Wick rotation, one is left with essentially the same problem: regulator fields for scalar fields obey Fermi statistics, and those for spinor fields obey Bose statistics. This violation of the spin-statistics theorem means that the Hamiltonian cannot be a positive definite operator, again implying the existence of negative energy states.

Solomon has argued that for QFT to be gauge invariant the Schwinger term must vanish, and Schwinger[8] long ago showed that for this to be the case the vacuum state cannot be the state with the lowest free field energy. The existence of non-zero Schwinger terms also impacts Lorentz invariance. Lev[9] has shown that if the Schwinger terms

---

[8] J. Schwinger, "Field theory commutators," *Phys. Rev. Lett.* **3** (1959), 296–297.
[9] F. M. Lev, "The problem of constructing the current operators in quantum field theory," arXiv: hep-th/9508158v1 (29 August 1995).

do not vanish the usual current operator $\hat{\mathcal{J}}^\mu(x)$, where $\mu = 0, 1, 2, 3$ and x is a point in Minkowski space, is not Lorentz invariant.

To begin with, however, it is important to understand the details of Schwinger's argument. Using Solomon's notation, the Schwinger term is given by

$$ST(\vec{y}, \vec{x}) = \left[\hat{\rho}(\vec{y}), \hat{\vec{J}}(\vec{x}) \right]. \tag{C.14}$$

Taking the divergence of the Schwinger term and using the relation

$$i[\hat{H}_0, \hat{\rho}(\vec{x})] = -\nabla \cdot \hat{\vec{J}}(\vec{x}), \tag{C.15}$$

where $\hat{H}_0$ is the free-field Hamiltonian when the electromagnetic four-potential vanishes, results in

$$\nabla_{\vec{x}} \cdot \left[\hat{\rho}(\vec{y}), \hat{\vec{J}}(\vec{x}) \right] = \left[\hat{\rho}(\vec{y}), \nabla \cdot \hat{\vec{J}}(\vec{x}) \right] = -i \left[\hat{\rho}(\vec{y}), [\hat{H}_0, \hat{\rho}(\vec{x})] \right]. \tag{C.16}$$

Expanding the commutator on the right-hand side of Eq. (C.16) yields the vacuum expectation value

$$i\nabla_{\vec{x}} \cdot \left\langle 0 \middle| \hat{\rho}(\vec{y}), \hat{\vec{J}}(\vec{x}) \middle| 0 \right\rangle$$
$$= -\langle 0|\hat{H}_0\hat{\rho}(\vec{x})\hat{\rho}(\vec{y})|0\rangle + \langle 0|\hat{\rho}(\vec{x})\hat{H}_0\hat{\rho}(\vec{y})|0\rangle$$
$$+\langle 0|\hat{\rho}(\vec{y})\hat{H}_0\hat{\rho}(\vec{x})|0\rangle - \langle 0|\hat{\rho}(\vec{y})\hat{\rho}(\vec{x})\hat{H}_0|0\rangle. \tag{C.17}$$

It is here that one makes the assumption that the vacuum is the lowest energy state. This is done by writing $\hat{H}_0|0\rangle = \langle 0|\hat{H}_0 = 0$. As a result, Eq. (C.17) may be written as

$$i\nabla_{\vec{x}} \cdot \left\langle 0 \middle| \hat{\rho}(\vec{y}), \hat{\vec{J}}(\vec{x}) \middle| 0 \right\rangle = \langle 0|\hat{\rho}(\vec{x})\hat{H}_0\hat{\rho}(\vec{y})|0\rangle + \langle 0|\hat{\rho}(\vec{y})\hat{H}_0\hat{\rho}(\vec{x})|0\rangle. \tag{C.18}$$

Multiply both sides of the last equation by $f(x)f(y)$ and integrate over x and y. The right-hand side of Eq. (C.18) becomes

$$\int d\vec{x}d\vec{y}\{\langle 0|f(\vec{x})\hat{\rho}(\vec{x})\hat{H}_0 f(\vec{y})\hat{\rho}(\vec{y})|0\rangle$$
$$+ \langle 0|f(\vec{y})\hat{\rho}(\vec{y})\hat{H}_0 f(\vec{x})\hat{\rho}(\vec{x})|0\rangle\}. \tag{C.19}$$

If Schwinger's "arbitrary linear functional of the charge density" is defined as

$$F = \int f(\vec{x})\hat{\rho}(\vec{x})\, d\vec{x} = \int f(\vec{y})\hat{\rho}(\vec{y})\, d\vec{y}, \tag{C.20}$$

the right-hand side of Eq. (C.18) becomes

$$
\begin{aligned}
2\langle 0|F\hat{H}_0 F|0\rangle &= 2\sum_{m,n} \langle 0|F|m\rangle\langle m|\hat{H}_0|n\rangle\langle n|F|0\rangle \\
&= 2\sum_n E_n \langle 0|F|n\rangle\langle n|F|0\rangle \\
&= 2\sum_n E_n |\langle 0|F|n\rangle|^2 > 0.
\end{aligned}
\tag{C.21}
$$

The left-hand side of Eq. (C.21) — essentially the form used by Schwinger — is here expanded to explicitly show the non-vanishing matrix elements between the vacuum and the other states of necessarily positive energy. This shows that if the vacuum is assumed to be the lowest energy state, the Schwinger term cannot vanish, and the theory is not gauge invariant. Solomon also shows the converse, that if the Schwinger term vanishes, then the vacuum is not the lowest energy state and the theory *is* gauge invariant.

For the sake of completeness, it is readily shown that the left-hand side of Eq. (C.18) becomes

$$i\int \nabla_{\vec{x}} \cdot \left\langle 0\Big|\hat{\rho}(\vec{y}), \vec{\hat{J}}(\vec{x})\Big|0\right\rangle f(\vec{x})f(\vec{y}) d\vec{x} d\vec{y} = i\langle 0|\partial_t F, F|0\rangle, \tag{C.22}$$

so that combining Eqs. (C.21) and (C.22) yields a somewhat more explicit form of the result given by Schwinger,

$$i\langle 0|\partial_t F, F|0\rangle = 2\sum_n E_n |\langle 0|F|n\rangle|^2 > 0. \tag{C.23}$$

Solomon's Redefinition of the Vacuum

In QFT, the invariant perturbation theory[10] leads to the Dyson chrono-logical operator $P(\mathcal{H}_I(\mathbf{x}_1) \ldots \mathcal{H}_I(\mathbf{x}_n))$, where $\mathcal{H}_I$ is the interaction Hamiltonian. The adjective "invariant" refers to the fact that time ordering in the Dyson series is Lorentz invariant if the $\mathcal{H}_I(\mathbf{x}_i)$ all commute at space-like separations. The chronological product can be expressed in a form where the virtual processes are explicitly repre-sented; that is, as a decomposition into normal products, where a Feynman graph can be used to represent each of the normal products. The vacuum state is empty (although vacuum fluctuations exist) and is the state of lowest energy. The theory is not, however, gauge invariant and a process of regularization and renormalization is used to make it so.

The virtue of the vacuum state to be defined in this section is that it allows QFT to be mathematically consistent in the sense that it becomes gauge invariant without the need for regularization. Because the theory is gauge invariant, the Schwinger term vanishes. The reason for this is that the definition of the vacuum is such that the usual vacuum state, $|0\rangle$, is no longer the state of minimum energy, and there exist states with negative energy. In the context of the cosmological constant problem, this means that even if one believes the summation of zero-point energies given in Eq. (C.5) is legitimate, it must be extended to the unoccupied negative energy states, leading to a significant cancellation in the summation.

Redefining the vacuum state may provide a more elegant means of resolving both the gauge invariance difficulties of QFT and the cos-mological constant problem. The specific definition of the vacuum state given below serves, at a minimum, as an heuristic example of this approach.

Because of its intuitive nature, hole theory will be used to introduce Solomon's redefinition of the vacuum in quantum electrodynamics. He has also implemented this redefinition in the context of QFT, and that will also be described here. It may therefore be useful to recall how the transition to QFT is made.

---

[10] S. S. Schweber, H. A. Bethe, and F. de Hoffman, *Mesons and Fields Vol. I Fields* (Row, Peterson & Co., White Plains, NY, 1955), p. 192.

In terms of positive and negative energy electrons, second quantization gives the Hamiltonian for a free electron,

$$H_0 = \sum_{ps} E_p\left(a_{ps}^\dagger a_{ps} - b_{ps}^\dagger b_{ps}\right),$$

(C.24)

where $a_{ps}^\dagger$ is the creation operator for positive energy electrons and $b_{ps}^\dagger$ creates negative energy electrons. In terms of the number operator N,

$$H_0 = \sum_{ps} E_p\left(N_{ps}^+ - N_{ps}^-\right),$$

(C.25)

where $E_p > 0$ and N_{ps}^- is the number of negative energy electrons.

Dirac's hole theory rescales the energy so that

$$H_0' = \sum_{ps} E_p\left(N_{ps}^+ + (1 - N_{ps}^-)\right).$$

(C.26)

If N^- (for a given p and s) vanishes — i.e., when a negative-energy electron is missing, then N^+ increases by one and we see a positive energy electron and a hole in the negative energy continuum, which is interpreted as a positron.

For the hole theory approach to dealing with negative energies to work, it is essential that the particles filling the negative energy states obey the Pauli exclusion principle. The technique would not work for bosons associated with the Klein–Gordon equation. While the particles filling the negative energy states must not produce an electric field or contribute to the total charge, energy, or momentum, they nevertheless must respond to an external field.

The transition to QFT is accomplished by setting

$$a_{ps} = c_{ps} \quad \text{and} \quad b_{ps} = d_{ps}^\dagger$$

(C.27)

so that destroying a negative energy electron is equivalent to creating a positron. The Dirac negative energy sea vanishes since electrons and positrons are treated as separate entities. Provided the b's satisfy the anticommutation relations

$$\{b_{ps}^\dagger, b_{p's'}\} = \delta_{ss'}\delta_{pp'} , \tag{C.28}$$

we have that

$$N_{ps}^{\text{positrons}} = 1 - N_{\bar{p}s}^- = 1 - b_{ps}^\dagger b_{ps} = b_{ps}b_{ps}^\dagger = d_{ps}^\dagger d_{ps}$$
$$N_{ps}^{\text{electrons}} = N_{ps}^\dagger = a_{ps}^\dagger a_{ps} = c_{ps}^\dagger c_{ps} . \tag{C.29}$$

As a result,

$$N_{ps}^{\text{positrons}}|0> = N_{ps}^{\text{electrons}}|0> = 0 , \tag{C.30}$$

since $|0\rangle$ contains neither positrons or electrons. The free-electron Hamiltonian in terms of the number operator, now sums over both electrons and positrons and is given by

$$H_o = \sum_{ps} E_p \left(N_{ps}^{\text{positrons}} + N_{ps}^{\text{electrons}} \right) . \tag{C.31}$$

Thus, one can readily move between hole theory and QFT although, as Solomon has shown, one sometimes obtains differing results because of the way the vacua are defined.

With reference to Fig. 1, Solomon defines a state vector $|0, \Delta E_W\rangle$ as the state where a band of negative energy states extending from $-m$ to $-(m + \Delta E_W)$ is occupied by a single particle (the exclusion principle holds); all other single particle states are unoccupied.

The vacuum state is defined by Solomon as $|0, \Delta E_W \to \infty\rangle$, where it is important to understand that the limit $\Delta E_W \to \infty$ means that ΔE_W goes to an arbitrarily large but finite number. If ΔE_W were set equal to infinity one would be reproducing the Dirac vacuum. This definition of the vacuum allows transitions from the occupied negative energy states within the band to those beneath the band, thereby making the Schwinger term vanish, and the theory is consequently gauge invariant.

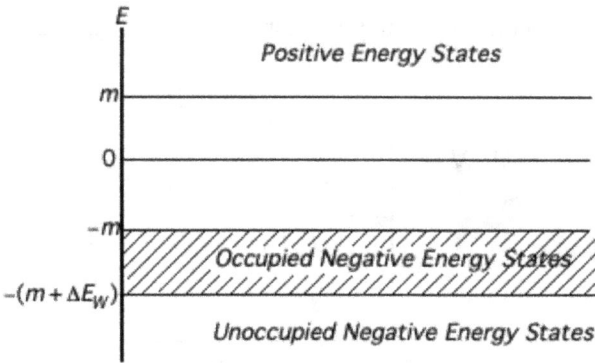

Fig. C.1. The state vector $|0, \Delta E_W\rangle$. Only the band of negative energy states extending from $-m$ to $-(m + \Delta E_W)$ is occupied.

The vacuum can be similarly redefined in the context of QFT. With reference to Eq. (C.27), the creation and destruction operators in QFT obey

$$c_{ps}|0\rangle = d_{ps}|0\rangle >\, = 0 \,, \tag{C.32}$$

while new states are created by $c_{ps}^{\dagger}$ and $d_{ps}^{\dagger}$ operating on the vacuum state $|0\rangle$. Solomon[11] redefines the vacuum to be $|0_R\rangle$, as $R \to \infty$ with the following restrictions:

$$c_{ps}|0_R\rangle = 0, \forall p \,,$$
$$d_{ps}|0_R\rangle = 0, |p| < R \,, \tag{C.33}$$
$$d_{ps}^{\dagger}|0_R\rangle = 0, |p| > R \,.$$

New states are created by

$$c_{ps}^{\dagger}|0_R\rangle, \forall p \,,$$
$$d_{ps}|0_R\rangle, |p| > R \,, \tag{C.34}$$
$$d_{ps}^{\dagger}|0_R\rangle, |p| < R \,.$$

---

[11] D. Solomon, "Another look at the problem of gauge invariance in QFT," arXiv: 0708.2888 (12 September 2007, V2).

Graphically, these conditions may be displayed as in Fig. C.2.

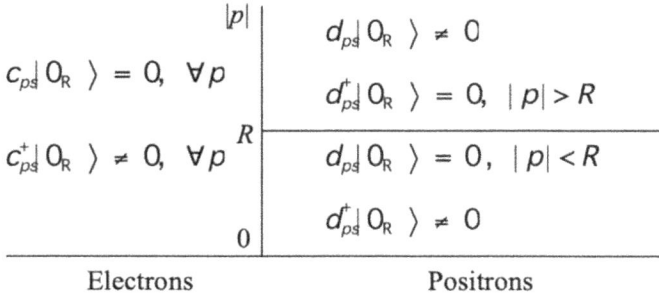

$$
\begin{array}{c|c}
 & |p| \\
 & \\
c_{ps}|0_R\,\rangle = 0,\ \forall p & d_{ps}|0_R\,\rangle \neq 0 \\
 & \\
 & d_{ps}^{\dagger}|0_R\,\rangle = 0,\ |p| > R \\
c_{ps}^{\dagger}|0_R\,\rangle \neq 0,\ \forall p & R \\
 & d_{ps}|0_R\,\rangle = 0,\ |p| < R \\
 & \\
 & 0 \quad d_{ps}^{\dagger}|0_R\,\rangle \neq 0 \\
\hline
\text{Electrons} & \text{Positrons}
\end{array}
$$

Fig. C.2. Solomon's redefinition of the vacuum in QFT. Note that $d_{ps}|0_R\rangle$, $|p| > R$ creates states with energy less than $|0_R\rangle$.

As $R \to \infty$ the usual relations of Eq. (C.32) are recovered. Solomon shows that the Schwinger term vanishes using this redefinition of the vacuum so that the theory is gauge invariant. Again, to achieve this it was necessary to introduce negative energy states.

The key question to be addressed, with this or any redefinition of the vacuum that allows negative-energy states, is stability. First, can positive energy particles be scattered into unoccupied negative energy states? And second, is the vacuum catastrophically unstable, in the sense that the transition rate from occupied negative energy states within the band shown in Fig. 1 to those below the band, so great that the band vanishes essentially instantaneously?

With regard to the first question, the idea here is to make ΔE_W large enough to make the probability of such transitions negligibly small. The answer to the second question is still open, but Solomon has determined that for a field theory of non-interacting zero-mass fermions in the presence of a classical electromagnetic field in a two-dimensional space-time, the vacuum is stable. It remains to show that this remains the case for four-dimensional spacetime.

Origin of the Inconsistencies in QFT and Summary

The problem of maintaining gauge invariance in QFT when interactions are present, and the difficulties with Schwinger terms discussed above, stem from the now well-known fact that the underlying assumptions of QFT are inconsistent. The essence is contained in Haag's theorem,[12] which is concerned with the interaction picture that forms the basis for perturbation theory. Haag's theorem is important not least for the fact that it identifies the reason regularization and renormalization are needed in QFT; that is, the underlying assumptions of relativistic QFT are inconsistent in the context of interacting systems.

In QFT, relativistic transformations between states are governed by the continuous unitary representations of the inhomogeneous group SL(2,C) — essentially the complex Poincaré group. One might anticipate that when interactions are present, the unitarity condition might be violated. Indeed, Haag's theorem states, in essence, that if ϕ_0 and ϕ are field operators defined respectively in Hilbert spaces $\mathscr{H}_0$ and $\mathscr{H}$, with vacua $|0\rangle_0$ and $|0\rangle$, and if ϕ_0 is a free field of mass m, then a unitary transformation between ϕ_0 and ϕ exists only if ϕ is also a free field of mass m. Another way of putting this is that if the interaction picture is well defined, it necessarily describes a free field.

The assumption that the vacuum state is the minimum energy state, invariant under a unitary transformation, is one of the fundamental assumptions of QFT. But it is now known that the physical vacuum state is not simple and must allow for spontaneous symmetry breaking and a host of other properties, so that the real vacuum bears little relation to the vacuum state of axiomatic QFT. Nevertheless, even if the latter type of vacuum is assumed, the violation of the unitarity condition in the presence of interactions opens up the possibility that the spectral condition, which limits momenta to being within or on the forward light cone, may also be violated thereby allowing negative energy states.

---

[12] R. F. Streater and A.S. Wightman, *PCT, Spin & Statistics, and all That* (W. A. Benjamin, Inc., New York, 1964), Sect. 4–5. See also, P. Roman, *Introduction to Quantum Field Theory* (John Wiley & Sons, Inc., New York, 1968), p. 388.

Of course, the way QFT gets around the formal weakness of using the interaction picture is to regularize the singular field functions that appear in the perturbation series followed by renormalization. There is nothing wrong with this approach from a pragmatic point of view, and it works exceptionally well in practice. The reason one may want to look at other approaches, such as redefining the vacuum state and the role of negative energies, is that it may lead to insights into the nature of the vacuum itself, and help resolve the outstanding cosmological constant problem, and could offer a new approach and possibly an alternative to the process of regularization and renormalization.

If it is indeed possible to use the negative energy spectrum to cancel much of the vacuum energy density of Eq. (C.5), there is now observational evidence pointing to what the residual vacuum energy density must be. This comes from the observational data of the Supernova Cosmology Project.[13]

---

[13] S. Perlmutter, "Supernovae, dark energy, and the accelerating universe," *Physics Today* (April 2003).